White Lion

BACK TO THE WILD

WHITE LION

BACK TO THE WILD

This is the story of the first white lions
to be released back into the wild.
But it's not just the story of the white lions –
it also tells of how a game reserve was created
from overgrazed sheep and cattle farms
to become a free roaming, Big Five wildlife
reserve, in an area that had not seen big game
for over 250 years.

ANDREW SCHOFIELD

Book production by Quickfox Publishing
on behalf of Andrew Schofield
PO Box 12028 Mill Street 8010
Cape Town, South Africa
www.quickfox.co.za

ISBN 978-0-620-51478-1

Photo credits:

All photographs were taken by Andrew Schofield except for the photographs on top of page 132 the two females by Rudolph van Niekerk and on the left hand bottom of page 132 author with sedated lion by Yvonne Mulcahy and on page 262 of author putting a collar on a cheetah by Malcolm Truman.

Edited by: Lucy Carreau and Daphne Stokoe
Proofread by: Maddison Haut Parleur
Book cover design by Andrew Schofield
Book design & typesetting by Vanessa Wilson
Printed by Quickfox Publishing

Author contact details:

Andrew Schofield
andschofield@gmail.com
www.andrewschofield.co.za

THIS BOOK IS FOR
SIMMY,
WITH LOVE

Thank you so much for all of the
amazing times we have shared.
I look forward to many more.

ACKNOWLEDGEMENTS

To my Mother, Mary, for a lifetime of encouragement and for nurturing a love of books. To Janet and Adrian my incredible children, with much love – I am so very proud of your wonderful lives. You inspire me. Paul Vorster and Karlien Heunis – you have changed many lives with your commitment to conservation and to a new generation of youngsters and I salute you. Dr Ian Player for being an incredible example to us all and for giving me such good advice regarding this book. Dr. Gaston Savoi and his wife Gilda, sons Philippe, Rodrigo and Guillermo, daughter Carolina and Alicia. Gaston, a special salute to you for making your vision and the improbable come true. Marissa, Ryno, Kevin & Karen, Izak, Koos and Willem, Jan, Barnard & Cornelia, Theresa and the rest of the team on Sanbona. To Veronique and Virginie, Yvonne, Suzy, Rachel, Nico, Rudi, Malcolm, Michaela and Ian, Kelly, Alouise, Scott, Andrew and all of the other students, rangers and wildlife team at Sanbona. A special mention of John and Janie Taylor and Lindsay and Kate Birch. Keir and Liesl for three days in the mountains. Adrian and Shirley Anne Gardiner and their family for their commitment to conservation. Joe and Johan and the team at Shamwari and Paul Gardiner. Nick Clegg at Dubai World.

A very special thank you to Simmy for keeping my grammar and spelling in check.

Due to the greyness of my hair and the simultaneous memory loss that has accompanied my ageing process, I have probably made some rather simple errors in the book and for this I apologise profusely. With the help of many people I have attempted to keep as accurate a timeline as possible. However, some of the incidents that took place are not in their exact sequence.

Although this is a story about lions – specifically white lions – it is also the story of the creation of a reserve, the incredible assortment of people and animals and all of the ups and downs that go with it. My special thanks and love go to the two pioneers, Jabulani and Queen. You will be with me for the rest of my life.

In the end, we will conserve only what we love.

We will love only what we understand.

We will understand only what we have been taught.

Baba Dioum

CONTENTS

PROLOGUE

Lions used to roam over most parts of the world including Africa, North America, Europe and Asia. Today they are restricted to central and southern Africa and to a small area of western India. Sadly, there are only about 15,000 lions in the wild today whereas that number was closer to 50,000 only 12 years ago. The 'King of the Beasts' is in big trouble, mostly from loss of habitat, with disease also a contributing factor.

The subspecies of lion, *Panthera leo persica*, which was found from India to the Middle East, can today only be found in the Gir National Park and Forest sanctuary which is situated in Gujarat, in western India. There are only about a stupefying 400 left in the world. There are two subspecies of lion as far as genetics go: *Pantherinae Panthera leo leo* – the African lion – and *Pantherinae Panthera leo persica* – the Asian lion. Narasimha, the 'man-lion,' is described as an incarnation of Vishnu and is worshipped as the 'Lion God,' so the Asian lion which was commonly found throughout most of India in ancient times is considered sacred by all Hindus. The very common surname 'Singh' is an ancient Indian name meaning 'lion.'

Stories and sightings of white lions have been around for centuries in Africa, but were often dismissed as superstition. White lions have been a part of San and African folklore from the beginning of time. From central Africa all the way to the Cape in South Africa, tales of them abound. Tragically, there are very few in the wild today, probably less than five at any given time. However, there are at present more than 300 in captivity.

The first documented sightings of white lions by Europeans took place in 1928. The first sighting in the Timbavati was in the early 1940s by Joyce Mostert, whose family owned large tracts of land in the area. During March 1959, 12 adult lions with two white cubs were seen near Tshokwane in the Kruger National Park.

In 1975 a litter containing two white cubs was found adjacent to Kruger National Park by researcher Chris McBride. Their story is documented in his book *The White Lions of Timbavati.* There were three cubs, two male and one female, of which one male and the female were white. McBride decided to have the three captured and they were kept in zoos and bred. The tawny male was even manipulated to breed with his own daughter, which produced a white offspring. It is from these three that most of the white lions in captivity today originate. Since the removal of those three, only a few white or pale cubs have been known to be born in the wild.

In 1979 a white lion was observed in the Umfolozi Game Reserve in Zululand. One female lived for several years and was often seen hunting with her pride. I have recently received a sighting on the border of Botswana and the Caprivi Strip. At present, all of the white lions in captivity come from the greater Kruger Park area. This doesn't mean that white lions came exclusively from this area, but this is where the most sightings in recent history have taken place.

The genes for producing white lions in the wild have been compromised by every removal of a white lion from the wild. So to produce a white lion, most breeders are forced to inbreeding. In certain breeding farms and zoos this has led to what is known as inbreeding depression, which can lead to infertility and other complications.

The Klein Karoo (Klein means small in Afrikaans and Karoo means dry in the San language) is one of the most beautiful places on this planet. Its harsh, arid environment is home to the succulent Karoo biome. The Karoo is one of the world's most extraordinary arid zones – it is a biodiversity 'hotspot.' The number of plant species in the region is unparalleled anywhere else in the world for an area its size, with at least 40 per cent of these species being endemic. Most are succulents, plants with thick, fleshy tissue that can store water. One-third of the world's 10,000 species of succulent plants grow in this area.

In 1982 I bought a farm there and over a period of years turned it from an overgrazed sheep and goat farm to a nature reserve. I soon learnt that the environment was not only incredibly harsh, but also very vulnerable to man's stupidity. The area had been overgrazed by sheep farmers for so long that some of the plant species had become locally extinct.

If any area in the world was designed to carry game in small numbers, this was it. The mouths of most of the antelope species are pointed and are almost able to 'prune' the plant, which is so much better than the way sheep and goats devour the whole plant. If you've ever watched a herd of antelope grazing you would have noticed that they're always on the move, taking a small bite from a plant as it passes and then moving on to the next one, nibbling here and there without decimating the plant. Then look at a flock of sheep – they'll stay in one area all day and eat each plant down to the root.

On my farm, I only introduced species that had historically occurred in the region; they did very well and produced offspring. I have always been of the opinion that a translocation is successful when the offspring born in the new area produce young of their own.

Black Eagle farm, as we called it, flourished. There were wonderful resources like natural springs and rocky outcrops where the San/Bushmen people had left beautiful rock paintings for us to admire and protect. It was also during these years that I started my formal studies. Having been hopeless at school, I'd finally found my niche and became a serious student and later went on to lecture at the university. My children, Janet and Adrian, grew up spending a lot of time in the wild, experiencing the outdoor environment with wonderful abandon. They have both become conservation-minded people who still treasure the memories that were made on the farm. Black Eagle farm also awoke in me an awareness of birds of prey (raptors), and I started a research project on pale chanting goshawks in the Klein Karoo which is still ongoing, nearly thirty years later. I also got the opportunity to learn from some incredibly talented and dedicated people about the raptors of southern Africa and the critical role they play in the environment.

In 1997 I was approached to create a wildlife reserve due south of our farm and after a family consultation, I agreed.

The Cape wildlife reserve started off when the Linton group began buying up farms north of Barrydale, eventually comprising a total of 28,000 hectares made up of six farms. Two really good people, John Taylor and Lindsay Birch, started the initial phases of creating what is now one of the finest nature reserves in southern Africa. They asked me to start the impact assessments and get the reserve on the go. I still to this day, see both of them every now and then and value their friendship. I still wonder what miracles might have happened if they'd got the support they needed to finish the project.

In those early days I had a very small team helping me. We had to initiate all of the ecological studies and do the necessary work to create a nature reserve ourselves. It was a very interesting time, going through all the stages one has to before opening a reserve. The crazy part about this is that you can buy any farm anywhere and put as many sheep, cattle and goats on it as you like, but if you want to start a game farm, the authorities put up mighty hurdles that you have to jump over. One of the requirements was meeting with the local community. It was incredible how much support we had from some of the local farmers, but what was really interesting to me was that the local tourist bureau took an almost hostile attitude towards the creation of the reserve. To this day I'm still not sure if it was a personality clash, or just a jealousy issue; but thankfully that changed at a later stage.

One aspect that was very important to us all was that we tried to use local labour to do as much of the work as possible. This brought about an incident that was to make me very wary of some people's motives. Three local businessmen decided to create a forum of sorts and declared themselves the only people with whom we could negotiate when taking on labour. They asked for a meeting and we met at the main gate to the reserve. After listening to them explain how they had created this organisation and being told I was now effectively over a barrel as to who I was able to employ, I politely asked them when it was that they had started the operation. I was told that they had started it in the past week and that they were up and running already.

I then explained that during this initial phase, our staffing requirements were minimal. Only when the reserve's lodges were going to be built would there be work for builders and catering staff. At this point one member physically leapt into the air and started shouting and performing to the point that I thought he might have an apoplectic fit. I was very glad that Kobus, my assistant, was with me because he came from strong farming stock and stood six foot two in his socks. This man was so beside himself that he began frothing at the mouth and became totally incoherent with rage. Kobus and I were literally gobsmacked at this tirade and looked at his two partners for guidance.

They just shrugged their shoulders. I called for an end to the meeting and requested that we meet in town once the gentleman had had time to calm down and we could speak in a more civil manner.

Sadly, the meeting didn't end on a happy note, with the man demanding to know the new owners' names and address. This individual then got in his car and, without an appointment, drove the 250 km to Cape Town and went on the rampage with John and Lindsay. John phoned me later and requested me kindly not to send any more crazy people down to head office.

I was later to find out that the three had seen an opportunity to make a fast buck, not realising that it takes a long time to create a reserve. Speaking to one of the locals I explained the situation and he started laughing, telling me that the nickname of the chap who had erupted was 'Kortsluiting' which is the Afrikaans word for 'short circuit.' We had one more visit from this trio and thereafter they left us alone.

John and I went on many reconnaissance trips in the area looking for possible farms to add on to the reserve. He made an appointment for us with one of the farm owners and on our arrival we were met by two couples drifting around in their underwear. We found it a little strange that they didn't even bother to put on clothes for our visit, and we had to do all of the negotiations with people running around in their undies. When the deal went through we named the place 'Onderbroeks' which is the Afrikaans term for underpants. I am delighted to say that on a recent visit to the reserve, I looked at the maps and found that the name had been retained. It gave me much enjoyment to explain to people how it came about.

On one of our exploratory walks I found out why John always wanted me to walk in front of him. Walking comfortably along, I suddenly stepped on a puff adder, and my reaction took me into a vertical leap which ended with me in a mangled heap on the ground, breathing ten to the dozen. Fortunately, one end of the snake was coiled around a plant and it couldn't whip back and bite me!

Over the next year, using local labour, we removed close on 270 km of internal fencing and hundreds of tons of scrap and rubbish that had accumulated on these farms over the years. One farm in particular, at the foot of the Warmwaterberg, was so full of rubbish that it took months to get rid of all of the debris. It's difficult to comprehend how the previous owners had desecrated this beautiful environment with so much litter, rubbish and scrap. Just the concept of transporting so much junk to this isolated spot is amazing. After a massive clean-up operation, things eventually improved to such an extent that the place became habitable and pleasant; we called it Tilney Lodge. Strangely enough, the farm right next door, called Hoogfontein, was the most pristine in the district. The veld was in excellent condition and the environment in a very healthy state.

Early one morning the team and I were on our hands and knees around a vegetation patch when we all heard someone say, 'Goeie more,' which means good morning in Afrikaans. We looked around, and then at each other, trying to identify the person who had just greeted us. We then got a slightly more insistent 'Goeie more' and there, about ten metres above us, was a raven gliding around and being ever so friendly. We all fell about laughing when we realised that we were being greeted by a bird! It must have been reared by somebody in the district and learned a few words.

The introduction of game began in earnest and the reserve was taking shape nicely when we had a very rude awakening. The investors put a stop to all our work and we became a caretaker operation for a year while the bank decided what they were going to do. This was a very difficult time for us all because we had no commitment from the bank, virtually no budget to do anything, and all of the staff feared for their livelihoods. During this time I learned some very hard reality lessons – till then I had only dealt with bank managers for small amounts of money. This was hardnosed business and I was dealing with bankers who quite frankly didn't give a damn about the reserve or the environment, or the people for that matter – just their cash flow. Help came when a new consortium consisting of a bank and Shamwari owner, Adrian Gardiner, took over the reins. I was asked to stay on and help get the reserve back on track. Having never worked for a corporate before, it was interesting to see how a big organisation works, and all of the politics that are associated with it.

Adrian renamed the reserve Sanbona, and things then happened at whirlwind speed. Extra farms were purchased in the north and lodges were built. Eventually the reserve grew to 55,000 hectares and opened its doors to the first guests. In the beginning, as the wildlife manager, my team consisted of Ryno Erasmus, my second-in-command, and our assistant, Theresa, a delightful young lady who had grown up on one of the farms that the group had acquired. A small team of helpers joined us: Cecil, Barnard, Koos and Izak who were all farm workers from surrounding farms or had been taken on when the owners of the farms sold to the group. None of them had any wildlife background at all but they soon grasped what we were trying to do and embraced the idea of a nature reserve.

Soon after, Kevin and Karen joined the team; Kevin was the student co-ordinator and Karen was in charge of the office for a short while. But they soon joined the hospitality department and ran Tilney Lodge, the first lodge for guests on the reserve. The infrastructure was growing all the time and a workshop was built, then a student accommodation facility which we called Gondwana. The students were to play a pivotal role in the wildlife department's function, and I was given the go-ahead to recruit two student co-ordinators. Karlien Heunis had done her practical year with me on the reserve as part of her degree and I knew that she would be the right person. Later, Paul Vorster joined the student team. Paul and Karlien were an amazing twosome, and there are young people all around the world who will never forget their time at Gondwana because of the selflessness of those two. The students were young folk, either wanting to do some practical work in conservation or to just enjoy a bush experience. Mostly female and British, the students certainly kept everyone on their toes. For many it was the first time away from home with the freedom to be responsible for their own actions. I had no idea that the Brits had such a capacity for alcohol! Many romances developed and some of those turned into marriages and children; for the rest, just very fond memories.

I really enjoyed working with the students as their enquiring minds would sometimes make me re-evaluate certain decisions. I am still in contact with many of them.

As with all endeavours, any project's success is reliant on a team, and I was blessed to have an incredibly committed group of people who put their hearts and souls into the reserve's success. So with this background to the reserve, when I write 'we,' it's the wildlife team I mean, made up of Karlien, Paul, Marissa, Ryno, Barnard, Izak, Koos and Willem, as well as a special lady who was in charge of the kitchen, Cornelia.

CHAPTER 1

INTRODUCTION

The average day in the life of a wildlife manager can vary from being snowed under by a pile of paperwork approaching the proportions and weight of an elephant, to being on the sharp end of a very grumpy leopard. In between all that is the incomparable privilege of being out in the wild among the animals and plants we love so much, observing the continuously changing pattern of life that makes every day a new beginning.

I started my wildlife journey as a young boy in Nigeria when I was entrusted with two bush babies to look after until they were ready for reintroduction back into the wild. I was blessed to have parents who allowed me the freedom to explore the forest and bush around Ikeja which, in those days, was a tiny suburb on the outskirts of Lagos near the airport. They had a wonderful friend named Paddy Courtney who was the local wildlife officer. He was a very patient and wise man who took the time to take a small boy under his wing and make the animals and plants of Africa exciting to learn about and observe.

Every one of his visits was eagerly anticipated as he always made some time to spend with me and we would walk in the dense forests around Lagos. He inspired me to look beyond the obvious and question everything – I must have driven the adults in my life insane.

Since then, over the years, I have been blessed to meet and work with some of the world's greatest conservation figures, and also to partake of wonderful conservation projects with them. Some of the highlights have been the reintroduction of various endangered species. In some of the areas I've worked in, the game had been wiped out more than 300 years ago, so to see an area change from sheep and cattle farms back into a natural, sustainable wildlife area is, for me, the realisation of my greatest wish for Africa. From elephants to small antelope and, of course, the big cats like lion, cheetah and leopard, I've had the privilege of working with most of them in one way or another. To this day I find it difficult to answer the question: which is my favourite animal? So many of the creatures I've had the honour of working with have left their particular mark on me in more ways than one. Whenever I'm around elephants they put me in a very relaxed state of mind. I have no idea what causes this, but my heart rate slows and my breathing becomes slow and deep and my whole being relaxes; and yet I'm still incredibly aware of everything around me. My wife tells me that I do exactly the same thing when I see whales.

Once while giving a martial eagle a final examination and ringing it prior to being released, I was at the same time explaining what I was doing to a group of students. While answering a question from one of them I looked up and got a talon right through my wrist, which returned my absolute and undivided attention back to the task at hand! I learnt once again that you can never let your concentration wander when dealing with a wild animal, however relaxed you think it might be.

Cheetahs have always been high on my preference list for many reasons and I shall go into that a bit later. But as I said earlier it has been the wonderful meetings with many very special people who have introduced me to a new way of thinking about subjects that I thought I knew a lot about. So, when one afternoon I got a telephone call asking me what I knew about white lions, a whole new chapter opened in my life together with a completely new understanding of certain ecological aspects. This was to bring to my life many different emotions, from anger and frustration to absolute joy and elation, plus wonderful memories that will remain with me for the rest of my life.

CHAPTER 2

THE PREPARATION

After a long night of reading and searching for as much information as I could, it became glaringly apparent that there was very little written knowledge about white lions. The one aspect that blatantly stood out was the hunting industry. Let me say at the outset that as a conservationist I have no dilemma whatsoever with ethical hunting. In fact, southern Africa has more game and areas for game now than in the early 1900s because of the value of the game for hunting and viewing. This is a reflection of the 'if it pays it stays' school of thought. Fortunately, large areas of privately-owned land are also used exclusively for game viewing and a massive tourist industry has built up around the game reserves.

This has also brought about employment for many thousands of people, plus the opportunity for small businesses to service the reserves. However, you will recall that I quoted 'ethical hunting.' The type of hunting that lions and white lions are subjected to called 'canned hunting,' is probably the lowest form of unethical behaviour known to man. The lions are bred for one purpose only – to enrich the pockets of the owners.

Lions are very easy to breed in captivity as they have a gestation period of about 110 days and they also have postpartum oestrus, so if they lose the cubs they can produce another litter within four months. Knowing this, the lion breeders remove the cubs soon after they're born and introduce a male and breeding takes place immediately. The lionesses effectively become factories and can produce three litters a year. In a natural environment, if the cubs survive, the lioness would only start to breed about two years later. At the lion breeding facilities the offspring are removed from the female and hand-reared. When they reach an age when they are of trophy value they are released into a small area where a so-called hunter then kills it. This is not hunting – it is just plain shooting, and any fool who can afford it, can do it. No fair chase or skill whatsoever is involved, and with the amount of money that wealthy Americans and Europeans are prepared to pay, this is a thriving industry. For the ignominious privilege of killing a lion in a cage, people have been known to pay up to $250,000 US dollars. I have nothing but absolute contempt for both the suppliers of these canned hunts and the dishonourable people who do the killing, which is exactly what it is and has no semblance to ethical hunting whatsoever.

So with all of this, I could only find out about white lions either in the hunting industry or in captivity in one form or another. From Las Vegas magical song and dance shows to zoos in France and South Africa and wildlife parks in the UK. Not one in the wild!

During the seventies two white lion cubs were found on the border of the Kruger National Park. They were monitored and when their condition deteriorated, they were captured and sent to the Johannesburg Zoo. These two lions were the founding gene pool of nearly all white lions in captivity today. If you mate two white lions, the offspring must be white.

It is important to understand exactly what a white lion is and, without too much scientific jargon, I'll endeavour to dispel some of the myths and nonsense that have been spouted over the years. A white lion is a lion that has a recessive gene. This recessive gene affects the hair colour and the colour of the pads (they are pinkish). Sometimes the nose is also a pink colour, and the eyes are sometimes a blue-green (in my observations of white lions in Africa, the UK and Europe, the males tend to have blue eyes and the females a green colour). The condition is known as leucism. The leucistic trait is due to the chinchilla mutation that inhibits the deposition of pigment along the hair shaft, restricting it to the tips.

The less pigment there is along the hair shaft, the paler the lion. As a result, white lions have a colour range from blonde through to a bright white. The white lion males have pale manes and tail tips instead of the dark brown or black that tawny lions have. White lions are not albinos. There have been noted cases of albinism in lions but this shows itself in different ways. Albinism is a congenital disorder that affects the pigment of the skin and is also associated with vision defects. I will refer to their hunting capabilities in depth at a later stage; suffice to say that lions are ambush hunters and are expert stalkers. They hunt mostly in the dark and they need to run down their prey very quickly. Rarely does the chase exceed 100 metres and most of the work is done by the pride females.

Doctor Gaston Savoi, from Uruguay, became a partner in the reserve. He was a very successful businessman and in his time had been a hunter, having successfully hunted many of the African species. It is an interesting fact that most hunters are actually very committed conservationists, doing extensive research on their quarry. While on a bush break, Gaston had the opportunity to read a book on white lions and after doing his research, found that there were none in the wild. Being the conservationist he is, he wanted to do something about it, and this is how the whole project started! Gaston had this vision to see white lions back in the wild. Over the years that I got to know him and his wonderful family, he was always very concerned about the welfare of the animals, and it was interesting listening to his observations – he has a very good knowledge of nature and sees things that very few people do. I have a great amount of respect for him, for his vision and commitment to seeing white lions back in the wild. Over the next few years he got some incredibly happy phone calls from me, but also some very upsetting ones, but he always maintained his dignified demeanour.

Knowing that there were no wild white lions we started researching methods of returning captive animals to the wild. This had been done before; however, the success rate was very low and the lions became problematic in the sense that they would trustingly approach humans. It's quite daunting having a lion run up to you at any time even when you know that the motive is to give you a big lick. The other problem is that they haven't been taught to hunt for themselves and the males are specifically at a disadvantage in these circumstances. Also, lions are the only social cats in the world, living in prides with sometimes as many as 15 other members. The order and safety of the pride is maintained by the dominant male of the pride, so to put a foreign lion into a pride will sign its death warrant. The team did an incredible amount of research and we spoke to as many people as possible. The discussions went around in circles for quite a while, the burning question being: how do we get the gene back into the wild without creating a breeding farm? The other aspect was that we would have to apply for a permit from the local conservation authorities with a motivation as to what our project intended.

With the knowledge that there were no white lions in the wild and that any lions which we could get hold of to start the process would be habituated to humans, we started working on a method that would be the safest and most humane, but also involve the least amount of human intervention possible. We decided not to even attempt to use the initial pair we hoped to get, but rather to do a breeding and work in such a way that the cubs would be the pioneers who would get the gene back into the wild. In short, this was going to be a long-term project in which we would have to work in lion terms and not human time. It was also vital that we were not going to become just another white lion breeding facility. The idea for the future was to get the white lion gene back into the wild so that, if one cub in ten thousand was born white, it would then have the chance of a natural life.

To keep wild animals on private property you need to have very good fencing, especially if you have the so-called

Big Five – elephant, buffalo, rhino, lion and leopard. Game-proof fencing is expensive and takes a long time to erect in places with hard soils. Electric fencing has improved greatly over the years. It is now very reliable and economical as most of the systems can be powered from photo voltaic solar panels. If someone wants to apply for a certificate to keep the Big Five, he needs to have the game-proof fences in place with five electric strands facing inwards. The five electric strands are offset at different heights up to the top strand at 2,4 metres. The bottom strand should be about 250 mm above the ground for digging animals like warthogs, and crawling antelope like gemsbok. The second would be at predator nose-height; the third at 750 mm for animals like hippo, rhino and buffalo. The fourth is at about 1,200 mm for the jumping species like eland and kudu, and also for young elephant that break the fence. The top strand is offset further than the others to stop elephants reaching over and pulling the fence poles down. This top one is also to stop giraffes from breaking the fence.

The word 'boma' is Swahili and means an enclosure, the fence normally made up of thorn tree branches with as many long, spiked thorns on it as possible, packed together to form an enclosure around their homes, cattle and sheep at night, protecting them from predators.

To have lions in a game reserve, the conservation departments insist on a minimum standard of fencing and these fences also have to have offset wires that are electrified. It is also good practice when translocating certain species to habituate them in a boma that is within an area of the reserve that represents the natural environment, before releasing them into the main reserve. (Rather like confining a baby in a playpen inside a room.) This allows the animals to get used to the new environment and the electric fence without stress. Normally when an animal has been shocked once they'll stay well away from the source of the pain. This also gives the wildlife team time to observe the animals for a while to ensure that no injuries occurred during the translocation before release.

The introduction boma for predators is normally rectangular in shape and about 50 to 70 metres long and about 30 metres wide. The narrow width makes it easier if an animal needs to be darted; most dart guns are effective up to about 30 metres. Due to the fact that we already had a pride of lions on the reserve it was also important to have an electric wire on the outside of the boma to stop the pride from attacking the new lions when they arrived.

The outline of the plan was quite simple, but would stress us all to breaking point! Like all pioneer endeavours, we would be doing things that had never been done before and would be under severe scrutiny by sceptics and nay sayers. It was also very important that whatever we did had to stand up to standard ecological practices and the lions' stress levels be kept to an absolute minimum – and in fact, create as normal a life as possible for them. We created a double boma, side by side, in an area that had suitable habitat for the lions – trees, grass and bush near fresh water and an interleading gate that we could open up and the two would then become one boma. Because we wanted to stop the lions from associating humans with being fed, we needed to create a feeding device that they would not connect with any human activity. This in itself was not an easy task. We tried all forms of devices, including a pulley on a cable that would swing the food over the fence of the boma. We scrapped that idea because of the possibility of injuring the lions. What we finally settled on was a chute into which a carcass would be released at different times, allowing it to fall into the boma. This would hopefully stop the human/food association.

Up to this point everything had gone smoothly, the enclosures being built to the Nature Conservation authorities' specifications with the electric lines in place. At this time we were also preparing a substantial document for the conservation authorities, stating exactly what we were going to do and why. At first it appeared that we would be denied a permit, as they thought it was going to be another lion breeding centre and queried what we would do with the offspring. Eventually, because of very good documentation and a working relationship that we had built up with the officers, they had enough confidence in us to give us the go ahead! With great relief on our part, we were told that the project had been approved by the conservation authorities and – in the fullness of time – a permit to import two white lions onto the reserve was issued.

CHAPTER 3

The Selection

At this time there were about 100 white lions in captivity in the world, and finding suitable breeding animals was a very difficult task indeed. Eventually, a lion breeder in the Mafikeng district was found to have two lions that were not directly related. However, we needed certain information for the programme to succeed; firstly that the female was not pregnant by another lion, and that the lions were healthy and didn't have any of the diseases that were, and still are, decimating the wild lion populations in Africa.

Bovine Tuberculosis (TB) is a killer in lions. The southern area of the Kruger National Park is seeing a decline in the lion population due to this disease. Lions are thought to contract TB from infected prey species, especially buffalo, being exposed to the bacterium that causes the disease when they eat infected animals. As lions often select an animal that looks easy to catch, buffalo weakened by TB are likely targets, and lions often scavenge on diseased carcasses. The first lion into the kill, usually a dominant male, will eat the choicest parts, such as the lungs, which are a main site for the TB lesions and bacteria.

Lions are also susceptible to canine distemper; in 1994 an epidemic swept through the Serengeti in East Africa. First symptoms are twitches in the front legs and facial spasms, which progress into intense seizures that completely incapacitate the lion. This unusual behaviour provokes attacks on them by pride members, so infected members became doubly infected by the disease, and injured by confused pride members. One third of the population was lost to this disease during that epidemic. It took a lot of research and study to find out where they had contracted the disease and eventually it emerged that the cattle herders' dogs had passed on the virus.

Feline immunodeficiency virus (FIV) does not appear to have any detectable health implications for the wild lion population. In fact, it would appear that wild lions have been living in harmony with the disease for many years. This disease was found in domestic cats in 1986. FIV produces a progression of symptoms over a period of a few years, causing immuno-suppression and leads to the eventual death of the cat. So although it does not appear to affect the wild population of lions, we can't discount the possibility that a lion in captivity may, as a result of the unnatural conditions and stress, develop a disease which would otherwise not affect a healthy wild population.

So, our selection criteria were that the lions were free of diseases and as distantly related as possible and that the female was not already pregnant. It was also very important to try and get the best representative specimens as possible as these lions were to create an added number to the gene pool.

I drove up to Mafikeng and met with Professor Henk Bertschinger from Onderstepoort, the veterinary faculty of the Pretoria University. We were going to take blood samples from the lions for the tests and to scan the female to check that she wasn't pregnant. It was also a requirement from Nature Conservation that the lions be micro-chipped. This is a small chip about the size of a grain of rice that has a unique code number, very similar to a bar code; it is inserted just under the skin using a strange-looking syringe with a fat needle, and when scanned the unique details are readable on a portable device. This technology is called radio frequency identification (RFID) and it doesn't need a battery because the micro- chip is activated when scanned. The reason for inserting the chip is so that if there is ever a dispute about the rightful owner of the lion, all it takes is a portable scanner to confirm the true identity and ownership of the lion.

This was my first visit to a lion breeding facility and the sights and sounds absolutely horrified me. However, when you deal with this kind of situation it's best not to show any emotion whatsoever. These are ruthless people and it's best to just keep to the programme and not open your mouth about what's going on around you. I would have the chance at a later stage to help the animals. There had recently been an expose on television about the canned lion hunting business in southern Africa. Knowing that most of the television programmes are purely there for sensationalism I hadn't taken the programme to heart till that moment. But I had just walked into what can only be described as hell on earth for any animal!

This was one of the most harrowing moments of my life, seeing hundreds of lions in cages of different sizes, the younger lions together in areas of about 20 x 20 metres and in some of them it was a mixed batch of young lions and young tigers, the adult females in their own small enclosures with a birthing shelter and the adult males in their own small caged area. What happened throughout the time I was there was that one lion in one sector of the cages would roar, and this would then trigger off a mass roaring by most of the adults, both male and female. As loud as it was, it was terribly upsetting knowing that the vocalisation was a territorial one, and that it could never achieve anything as the lions were just too close to each other. Lion territories in the wild are vast, and roaring is one of the ways that warns off intruders and marks the territory of a pride.

The owner escorted us everywhere and made sure we didn't wander off to have a look around, and when I brought out a camera to film the procedure he got very angry. The only reason that I got away with it was that I said I was told to do it by the owners of the reserve and that if I didn't have photos of the lions they wouldn't continue with the transaction. This wasn't entirely true but after reassuring him that I wouldn't film the other lions and tigers and bear, he relented.

Henk had brought the scanning machine with him and we darted the female first. This was my first opportunity to do an examination of a white lion up close, and I took my time to do an in-depth study of her while Henk did the scan. She appeared to be a four- to five-year-old, and she was in good condition. The drug used in the dart to anaesthetise lions is called Zoletil and one of the effects is that the lion loses the ability to close its eyes, so it is good practice to put a gel into the eye to keep it moist and then to put a cloth over the eyes. As I was applying the gel it gave me the time to look closely into the eye of the female – it had a slight green tinge to it but the background had speckled brown spots. This was very interesting indeed, very different from all of the lions I'd worked with in the past. The scan was completed and the blood samples taken and the female was returned to her three by ten metre cage to come around from the anaesthesia. When Zoletil is used, there is no antidote and the animal wakes up when the effect of the drug wears off. Depending on the initial dose, this is normally after about 30 minutes but the full effects are felt for a lot longer, so it is important that the animal is monitored during this time as they are disorientated and vulnerable.

We moved on to where the male was and took his blood samples. While we were busy working on him the breeder brought the two lions' files for me to look at. The male looked younger than two years of age, however he was in good condition although overweight. This is a problem with hand-reared captive lions; they get all of the titbits off the tables and they develop a line of fat along the centre of the belly extending from the chest to the groin. It shows itself in the form of a flap. While the blood samples were being taken it gave me the opportunity to examine him as well. What was interesting was that his eyes were blue, a beautiful aquamarine that looked quite out of place in a lion. Another strange thing for me was the colour and texture of the pads – they were pink and absolutely smooth. All other lions that I have examined have rough black pads and it was then that it dawned on me that these lions had never walked on the

African soil. They had lived their lives on concrete and their pads were as soft as a baby's bottom. I remember thinking what a difference they were going to experience when they arrived on the reserve and wondered just how long it would take for the pads to toughen up on the hard soils of the Klein Karoo.

The stud files confirmed our examination of the lions and it was interesting that the lions actually had been given names. On the reserve it was standard policy to use numbers for all of our lions and cheetahs, the number being the telemetry frequency that the animal was wearing. Because no predators had been in our area for hundreds of years and we didn't know how they would cope in this new environment, we fitted telemetry collars or inserts for study purposes and for monitoring their movements around the reserve. It is also a fact that male scientists tend to number their study animals whereas female scientists tend to name them. The names of the lions in the files were Jabulani and Queen.

I couldn't get the absolute horror of the breeding centre out of my mind and committed myself, there and then, to do anything in my power to stop this form of animal cruelty. I have subsequently been to as many of these facilities as possible and have exposed the bad ones for what they are, but this is very big business and some of the stakeholders don't mind using illegal and violent methods to keep their businesses going. Many have gone to extreme measures to ensure that nobody catches up with them.

On the long drive back to the reserve my thoughts were centred on the work ahead. The task had seemed so far away when it was in my mind and then on paper, but now that I had finally met the key role players it had become a visual reality! I had to stop off on the way home to collect some brown hyaenas in the small town of Britstown in the middle of the Karoo. John, the ecologist from Shamwari, brought them to me so that I didn't have to go out of my way on the way back to the reserve. It was around midnight when I got into the village and John had timed it perfectly, arriving at literally the same moment. We had rather a funny incident during the transfer.

A very large dog was walking down the street, barking its head off. As soon as it got closer and scented the hyaenas, it shot off with its tail between its legs, whimpering loudly as it went. We both burst out laughing at this instant reaction, but it just goes to show that even though that dog in all likelihood had never seen or smelt a hyaena before, it knew just from the scent that they weren't to be trifled with. The transfer of the animals went off without any further ado and we went on our respective ways again.

When I reported back to the team what had happened during the trip up to Mafikeng and what I had seen, my enthusiasm must have rubbed off and they embraced the project totally. From then on it was a wonderful collaboration between the student coordinators and their students and the wildlife team. We were all now getting the last bits and pieces in place and anxiously awaiting the arrival of the lions.

CHAPTER 4

The Arrival

It was a very quiet, warm Saturday afternoon when Jabulani and Queen arrived. Jabulani means 'come to bring happiness to everyone' in Zulu. He was certainly aptly named. Our team was more than able to transport the lions but the breeder had insisted on bringing the animals himself. The two bomas, side by side, were ready for their new occupants and the team on the reserve was ready for the challenge ahead. We released Queen into her area and she walked off giving us a good deep growl. She looked fine, even after having been in a crate on the back of a pickup for the 1,000 kilometre journey.

Jabulani walked out and lay down immediately. He had rubbed his eyebrows on the crate and they looked raw and painful. This was one of the last times that Jabulani would receive human contact without being darted. I went into the boma and put an antiseptic cream on the wounds and calmed him down. I checked the enclosure for the last time and walked out. We watched and waited till the two lions saw each other through the fence and fortunately there was no aggression whatsoever. We watched for a little while longer until both lions were relaxed, then left them. Ryno and our assistant, Marissa, were both unusually quiet for a while and then started bubbling; the excitement was tangible.

The fun and games started at my office. I am fairly well-travelled and have met people from all walks of life but this particular breeder could have made a boatload of sailors blush with his foul language. The transfer of the payment for the lions, which had in fact been done, didn't show up on his banking system because it was over the weekend; he didn't believe the new owners had made the payment and was incandescent with rage. My attempts at trying to assure him that the payment had been made fell on deaf ears and his shouting and swearing got quite venomous, ranting and raving at everybody and threatening to put the lions back in the crates and drive back to Mafikeng. It brought home to me the defective character traits of some of the lion breeders and canned hunting operators. After a while he calmed down and accepted the situation and left without Jabulani and Queen. And that, fortunately, was the last I ever heard from him. He never phoned to check up on his lions as he'd said he would.

That evening I inspected the bomas again to find that the water levels in both of the cribs had subsided and that the large portions of game meat that had been put into each enclosure before they arrived were decidedly smaller. Knowing that they had both had a drink and something to eat put me more at ease and I think we all had a good night's sleep.

I'm an early riser and would more often than not be in my office at four o'clock in the morning. I started doing this when I decided to further my studies and over the years it became a habit. Getting up early also gave me the opportunity to clear my desk of any correspondence and prepare for the morning meeting with the team. It also gave me some time to be out in the veld to see for myself what was going on.

That first morning when I arrived at the bomas I got a nasty shock. Queen was under a tree and I could see her clearly; she looked fine, but Jabulani was nowhere to be seen! I drove around twice but still couldn't see him, but there was no indication that the structure had been breached and also no lion spoor outside. I decided to park the vehicle further away and sit and watch for a while before I raised the alarm. As I drove away I looked in the rear-view mirror and there was Jabulani following the vehicle, and as soon as I stopped he disappeared again. This was his game but it also showed that if a white lion doesn't want to be seen, they won't. All of the stories that I'd heard about them not being able to hunt because they stick out were proved completely wrong. Jabulani had disappeared in a 30 by 70 metre enclosure. This game of hide and seek went on for many months to come – he would either be right out in the open or so well hidden you couldn't see him.

The day-to-day monitoring, feeding and watering was a combined effort but the student programme took the task to heart and did the work effectively and efficiently. During this extremely important time we had students from France, America, Australia and England. Two of the students would be waiting for me every morning at six o'clock to check on the lions. It's always a pleasure to work with people who are enthusiastic and Veronique from Paris and Suzy from California were always on time and on the drive would ask very pertinent questions that often would challenge me to go and do some extra research. Suzy is now a Park Ranger at Yellowstone National Park and Veronique runs a very successful independent ecological auditing firm in Paris.

The feeding was done from the chute which worked very well indeed. However, in retrospect, I'm convinced that Jabulani and Queen knew exactly where their meals came from. A very important decision was made at the outset, that all of the food that was given to Queen and Jabulani would be from the veld, so they had different kinds of game species on their menu like ostrich, kudu, wildebeest and springbok. Ryno is an excellent shot and it was mostly his task to catch their dinner every day. Depending on what animals were in the different bomas at the time, this was a very demanding aspect of our work.

About six months before the arrival of Jabulani and Queen we had acquired three wild lions for the reserve. Ryno was given the privileged task of collecting them with the vet, Johan. The night before he was to fly off we drove down to Cape Town and my wife Simmy and I took Paul and Ryno out for dinner at the Ocean Basket in Plumstead. Those of you who have eaten there will know that they bring side dishes of chilli and garlic. Well, between Paul and Ryno they polished off six bowls of raw garlic with Ryno consuming the king's share. The next morning I dropped Ryno at the small aircraft section of the airport to fly up and collect the lions in a Cessna Caravan. When Ryno saw the size of the plane he visibly paled and it was at that point that he told us that he was incredibly scared of flying. Paul was with us so we had a backup but Ryno was determined to bring the first lions in over 300 years to the Klein Karoo, and made the brave decision to go. I felt very sorry for the pilot because Ryno really hummed of second-hand garlic!

After they had collected the lions they flew down to the reserve where we were all ready to welcome them at our landing strip. After unloading the sleeping lions I had a chat with the pilot about the flight and his comment was that he didn't know what was worse, the smell of lion dung or Ryno's garlic breath.

Johan, the vet, installed an internal tracking device in the one female and we collared the other two as these were going to be the first free-roaming wild lions on the reserve and we needed to monitor them closely to see what area they would utilise and what they would hunt.

The day had turned cold as we lay the anaesthetised lions in the boma for their acclimatisation phase. One of the other aspects that affects the animals under anaesthetic is that they cannot maintain body temperature, so we put blankets over them and locked the gates making sure that the electric fence was off in case they stumbled into it while still under the effects of the anaesthetic; then we watched and waited for them to come round. Once they were stable we switched on the electric fence and went off to celebrate.

Early next morning as I was on my way down to the boma I stopped off at the student facility to find out how everybody was and as we were chatting, the lions started to roar. A beautiful dawn with lions roaring in the Klein Karoo for the first time in over 300 years. What a way to start the day! Little did we know that very soon we would also have white lions roaring on the reserve.

The first few days are always critical and the lions were monitored at least twice a day. The female that had the internal transponder was showing signs of distress and we called Johan back to the reserve. He wanted to dart her but because they had all been darted and sedated for a long time, he didn't want to dart the other two as well, as this could lead to complications. We decided that we would drive into the boma and dart only the female and protect her with the vehicle while she went under. What happens is that when they start to go under from the effects of the drug, they stumble and become disorientated and this behaviour can bring about aggression from the other lions.

The dart was fired and the lioness started to go under. The other two lions then started behaving aggressively towards her and we had to keep the vehicle between them. This was all going on in the confines of a space about 50 metres by 30, with Simmy standing anxiously at the gate so that she could open it when we needed to get out. Paul and Ryno are strong men but not even they could lift a lion onto a pick-up by themselves, so when the lioness went down we all jumped out of the vehicle and picked her up and put her in the back. During this time the other two were doing small charges and really being aggressive, so we all jumped in and I headed for the gate. As I got closer Simmy started opening the gate and it was at that critical moment that the other two lions decided to charge the pick-up carrying their mate.

The gate had to be opened wide so that the vehicle could get through and Simmy very bravely held it open for us to drive out at full tilt. She reacted with adrenalin-augmented reflexes and the three of us looked over our shoulders and watched in horror as the male lion charged at lightning speed towards the closing gate. After what felt like an eternity she got the gate closed but not latched as the lion slammed into the gate. There was a cloud of dust and the male smacked the ground with his paws in a whirling motion, sending dust and stones flying at Simmy, all the while growling and snarling at her. Trembling but triumphant, she finally got the metal ring over the hook of the gate and the lion retreated. We jumped out of the vehicle to congratulate her on facing down those lions. She was covered in bright orange dust and totally speechless. I was very proud of her fearless reaction to the situation because I can assure you it's not everyone who would react like that under such enormous pressure. Two lions coming at high speed and growling at the same time is enough to make many turn and run.

Johan inspected the lioness and found that some of the stitches from the operation to install the internal transponder had come undone. He repaired the problem and gave her a healthy dose of antibiotics. These were also good opportunities to get some of the veterinary students involved, and it was always good to see their reactions at the chance to work on a sleeping lion. Once the repairs had been done we put her into the next door boma to wake up. We monitored her for a while till she was back to good health and then opened the gate so that she could rejoin the other two.

Simmy has spent a lot of time in the bush with me and has been a very important part of the reserve even from a distance, but that day she impressed all of us with her sangfroid under serious stress.

enclosures and the three of us looked over our shoulders and watched in horror as the [illegible] charged at lightning speed towards the closing gate. After what felt like an eternity, she [illegible] the gate closed but not latched as the lion slammed into the gate. There was a cloud of dust, and the male smacked the ground with his paws in a [illegible] motion, sending dust and stones flying [illegible] growling [illegible] [illegible] of the gate [illegible] We [illegible] [illegible] tremendous pressure [illegible] and [illegible] to move to another section.

Johan inspected the fence and found that some of the [illegible] crossbeams [illegible] to get some of the [illegible] it was a good idea to see [illegible] Once the [illegible] had been done we put [illegible] for a while till [illegible] and then opened the gate so [illegible] could rejoin the other two.

[illegible] spent a lot of time in the bush with me and [illegible] part of [illegible] even from a distance [illegible] impressed [illegible]

CHAPTER 5

Getting Acquainted

The pioneer pride was released into the greater reserve after a six-week period. The female with the tracking collar headed for the hills and the male and the other female stayed together. This wasn't unusual at all; in fact, the acclimatisation process in the boma is also a bonding period but lions don't always bond and for many months that female stayed away from the other two and we had the devil of a time tracking her in the hills. We even resorted to doing an aerial tracking and found her in a deep valley where the tracking device couldn't receive the signal. She stayed away from the other two for about two months and then one day she appeared and the three became inseparable. These first three lions gave the team a very good learning period and good preparation time which benefited the white lions when they arrived.

Jabulani and Queen settled in very quickly and what was very encouraging was that after about a week they started lying down closer and closer to each other. The fence between them would hopefully stop any aggression flaring into violent attacks if that were going to happen.

My home on the reserve was about 5 kilometres as the crow flies from the bomas, and one morning, about two weeks after they'd arrived, I heard the most incredibly persistent roaring, and it sounded as though both the male and the female were vocalising. Thinking that it was the reserve pride, I grabbed the telemetry set and went tracking to see what was going on. As I drove down the road it became obvious I wouldn't need to switch on the telemetry as the roaring hadn't stopped, and it was coming from the direction of the bomas. As I rounded the last tree before the boma I saw the tawny male and female just outside of Jabulani's enclosure, roaring their heads off. If you've ever been close to a roaring lion you'll know that the sound is so loud it penetrates your whole being, makes you feel as if your insides are vibrating. I rushed off to check that the electric fencing system was working; this apparatus was situated about 200 metres away at the cheetah boma.

While I was checking the system the roaring stopped, and I hoped that the pride had moved off. The system was active and working properly so I went back and as I rounded the last tree, the male attacked Jabulani's boma; he was about five metres inside the fence. The noise was incredible and what I hadn't realised till then was that while the male was smacking the fence and growling, the female was growling at the same time, then she joined in the attack. In the safety of the bomas both Jabulani and Queen had joined in the roaring. The noise was intense as I drove forward and hooted, hoping to distract the pair and move them away. This was not to be and I needed to quickly find a solution because this could have fatal consequences. Even with the outside offset electric wire the male would attack the fence trying to get at Jabulani. As always our wildlife team jumped into action, and what we did was build a low level fence about two metres out from the boma fence. The team did this while the two reserve lions watched our every move from a shady spot about 200 metres away. We too kept an eye on them with enough vehicles so that the whole team could do a swift exit if necessary. The fence we built was just taller than the big male with just three electric strands and it had the desired results. The male returned after we'd finished and charged Jabulani, getting a really good shock and although he would continue to charge he never touched the fence again. These visits would occur again periodically and we always knew when they were around the bomas because of the tumultuous roaring. Jabulani and Queen could roar just as well as the reserve pride and when they all roared together it was truly extraordinary.

After about three weeks we turned off the electricity between Queen and Jabulani's boma. They would lie very close to each other and we wanted to encourage this and not have the negative association of a nasty shock. They very soon sensed when the electricity was off and they started greeting rubs through the fence. Lions greet each other by rubbing their necks and heads together – you'll often see this happen when a pride wakes up and heads off to find water or to start a hunt. Lions also have a funny habit of copying each other; if one yawns the whole pride is likely to yawn almost as if it's contagious. The same goes for grooming and even urinating and defecating. Jabulani and Queen were bonding and this was exactly what we needed to happen. There was another aspect of their behaviour that we noticed which was that even though they were being fed a very good and varied diet and had never seen another wild animal in their lives, they both reacted to any game walking near the bomas. Queen would even crouch down and do a short stalk in the direction of where the animal was. There was no shortage of passersby in the form of antelope and zebra. Even elephants would walk past the bomas if the wind was such that they didn't smell the lions.

It was during this time that we started to select an area where we could create a natural functioning system with all of the prey items, to put Jabulani and Queen in. The area had to support a lot of game, have suitable habitat for hunting and have water sources. The other aspect that was of great importance was that the area had to be secluded and away from human activity. An area of about 3 000 hectares in the far east of the reserve was selected. Fortunately two of the sides were already fenced so we created a triangular-shaped camp reducing the amount of fencing. We then started the introduction phase by bringing in truckloads of kudu, gemsbok, springbok and other prey species.

I had to go for a meeting in Port Elizabeth and left the reserve at about four am one morning. I was approaching Riversdale which is about 150 kilometres away over a high range of mountains, and was surprised to be able to pick up the chatter on the reserve's radio frequencies from such a distance. It was April the first and I couldn't resist the temptation! I contacted the reserve and told them to be careful of the lions close to Tilney Lodge, which was full of guests. Darryl, the maintenance specialist, realising what was going on added that he'd heard roaring in the area.

At this point I went out of the reception capabilities of the radios. On my return I was greeted by a very grumpy Ryno. I asked what was troubling him and he let me have it in no uncertain terms that April Fool's jokes were not appreciated at all! What had transpired was that he'd heard the call and thought I'd said that the lions were in the lodge itself. So, as quick as a flash he went off and shot a springbok so that he could entice the errant creatures away from the lodge. When he arrived to do the rescue he found Darryl laughing his head off and explaining that it was April Fool's day. This didn't go down well at all and I apologised profusely and promised I'd never do such a thing again. I even wrote an open letter of apology to all the staff and bosses, and included an apology to the family of the deceased springbok in it. Fortunately everybody regained their sense of humour and I became the butt of quite a few practical jokes myself.

Only one channel on our radio network communicated with our neighbours. To the north of the reserve was a farmer who was very conscientious about the fence line. I was driving towards his area when I heard him talking to his assistants on the radio and picked up that they were walking the fence line to inspect it for any damage. We had just introduced a black rhino to the reserve and all of the neighbours knew about this as it was the first free-roaming rhino to walk our area for centuries. I called him and asked exactly where he was, to which he asked with slight trepidation why I wanted to know. I crossed my fingers and said that the rhino was trotting down the fence line near his position and that he should be aware of it. By this stage the whole district was listening in and the conversation from his side became somewhat heated, with me saying that he needn't be scared because it was a very friendly rhino and that he and his team could call him and pat him like a puppy, to which he replied something along the lines of "Have you lost your f....ing mind?" To which I wished him a very happy April the first. Radios all around the district crackled to life with farmers from all around laughing and ribbing him and me. Because the local community knew he was a big tease, it was all taken in good spirit.

CHAPTER 6

TOGETHER

Jabulani and Queen were very relaxed around each other and would spend all of their days lying next to the fence of each other's boma, often rubbing heads and necks through the fence. The time was coming to introduce them to each other but this was really going to test our nerves to the limit. Anything can happen when you introduce lions – fighting can break out and if it did, Jabulani wouldn't stand a chance against Queen. But this whole project of getting the white lion gene back into the wild depended on these two. Their demeanour towards each other was very encouraging and the decision was made – let's open the gate! The day was decided on, and we had two loaded dart guns at the ready just in case things went awry.

With all twelve of the students standing behind me observing, a pole was used to push back the gate between the two bomas, and it didn't take long for Jabulani to work out what we'd done. He gingerly walked through the gate. Once on the other side, Queen started to approach him with her head down in an attitude that really didn't look promising. At this point we were all stressed to the limit with every negative possibility going through our minds, something along the lines of having to phone the bosses and tell them that there was a really big problem.

Nothing could have prepared us for what happened next, however. That head down attitude of Queen's was the biggest 'Come hither, Big Boy' I have ever seen. She literally threw herself underneath him and they started mating there and then right in front of us. You cannot imagine the relief, and of course with 12 young students giggling away behind us we also started laughing from sheer release from tension. And so it went on for five days, mating every twenty minutes or so. We couldn't have asked for a better outcome.

Introducing lions to each other is a very stressful undertaking, but this one had worked incredibly well and the prospect of the arrival of cubs now put us in a position to get the white lion gene back into the wild. In lions, only about one in four matings results in pregnancy; however, Queen was in exceptional health and had a very good, nutritious diet so the chances were good. It would now be a time to wait and see. When Jabulani arrived on the reserve he'd had a slight squint which didn't seem to affect his vision at all and he functioned quite normally. One of the results of the five days of mating was that his squint disappeared. I am not, for one moment, suggesting that having sex for five days will cure a squint – but it worked for Jabulani!

Lion biology is a very interesting subject and much has been written about it. There are very good reference books and papers available, but what we were doing here had never been done before so there was no book to learn from.

With so many people and authorities keeping an eye on us before we did anything, a lot of consultation was done and every aspect of any decision was thoroughly investigated before any action was taken. This was good in the sense that everybody had contributed to the project and as all of the team were highly experienced, some really ingenious aspects were incorporated into the project.

A new boma system had to be built in the selected white lion project area which was called Brakrivier. Brak is the Afrikaans word for brackish which means saline, and the river's water was incredibly salty, so the name was quite descriptive of the area. The boma that we were going to build had to satisfy quite a few aspects of lion biology so that when the cubs were born, there wouldn't be any visible human activity from their perspective, especially feeding.

Just before the female gives birth she leaves the pride and finds either a thicket or a cave to give birth in. She will then keep the cubs away from the pride for anything from four to eight weeks. The introduction to the pride of the new cubs is also a stressful time for the lioness as the males are not renowned for their gentleness and have been known to injure the little ones. With this in mind and the new area at Brakrivier being fenced, we sited the boma with a thicket in it so that Queen could hide there when it was time to give birth. The area also had to be big enough to create three enclosures alongside each other with interleading gates. In the two outside bomas we built small three by three-metre cages in the corner which would have the dual purpose of a wake-up area if any of the animals needed to be darted and a back-up method of feeding if the main system failed. The idea of having three bomas in a line was so that we could put Jabulani and Queen in the end ones where they couldn't see one another when Queen gave birth. However, they would still be close enough to vocalise with each other. And when the cubs got to about four weeks of age we could let Jabulani into the centre boma so that Queen and her cubs could interact with him through a fence for a while before we opened the gate and let him join his new family.

While all of this was going on, the reserve still had to continue running and all of us in the wildlife department were stretched to capacity. At the same time, Sanbona was being visited by many guests who stayed in the wonderful lodges of Tilney and Khanni. Between us, we were monitoring the lions, introducing new game onto the reserve and another first was about to take place.

The Klein Karoo was about to have a herd of elephants move in and sample her wares, also for the first time in about 300 years. This was a very big deal for me as I had studied the vegetation of the area for about 20 years and had come to the conclusion that the Klein Karoo was not going to get back to where it was in bygone times because some of the seeds from various plants and trees needed to travel through the alimentary canal of a mega herbivore, like a rhino or an elephant, before they could germinate. In fact, what I'd done previously was to harvest seed from various plants and put them in different acid solutions trying to mimic the digestive acids of various large herbivores. This worked incredibly well and I had great success with various species of trees.

Certain seeds have a waxy protective outer layer that won't break down by itself, and only by travelling through the alimentary canal of an animal such as an elephant is the protective layer removed by the acids in the gut. This is then deposited onto the ground in a ball of dung which, when the rains arrive, is the ideal medium for germination. I'm convinced that areas that have been deprived of the likes of elephants and rhinos have actually degenerated because certain species of plants that rely on them for regeneration have gone dormant. A classic example of this are two tree species: the Ghwarri tree, *Rhus undulata*; and the Boerboon tree, *Schotia afra*. While doing research and plant studies I noticed that there were no young trees of either of these species. In fact, on Sanbona, the youngest Boerboon tree was estimated to be 250 years old and this coincided with the demise of the last elephants in this area.

Fortunately, I had a very good working relationship with Brand van Breda at the Worcester Veld reserve which specialised in the restoration of the palatable plant species of the Klein Karoo. Oom Brand was one of those incredible people who had decades of first-hand practical knowledge and knew how to communicate this to every level of the community, from the scientific intellectual to the hardy sheep farm worker. He helped me create a small nursery with a view to re-establishing areas back to what they'd looked like hundreds of years ago. If you look at the vegetation of the Klein Karoo today it's hard to imagine that grass species were a major part of the landscape. Centuries of overgrazing have resulted in the habitat being altered to the extent that literally no grasses are to be found there. Brand found small areas of grasses that had escaped being annihilated and planted lands of Hartebeest Grass (*Chaetobromus dreageanus*) and Red seed grass (*Erharta calycina*), from which we harvested the seeds and planted them in areas around the reserve. It is interesting that 15 years later these species are spreading around the Klein Karoo again because they're not being decimated by sheep and goats, but nibbled on by springbok and eland.

I have many fixed point photography sites all over the area and over the years it's going to be interesting to monitor the changes that places like Sanbona will bring. In the archives there are accounts of the springbok migrations. In 1849 at Beaufort West Sir John Fraser documented that a herd of springbok took three days to pass the town! It's very difficult to imagine that there could have been that many in one herd. I don't think this spectacle will ever be seen again but to see a large herd out on the Karoo plains is very rewarding.

CHAPTER 7

ELEPHANTS

Before Adrian and the bank had taken over the reserve, John Taylor had dropped a bombshell by announcing in the newspapers that the reserve would have the Big Five. This meant that we would have to introduce buffalo, elephant, lion and rhino. There was a very small existing population of leopards which we had started to monitor. Sadly, they'd been persecuted to the point that they lived almost exclusively in the mountains. I knew from the start that we were going to face an uphill battle trying to get permits for the different species, specifically the elephant, buffalo and rhino. As in all pioneer endeavours, people put up barriers by being suspicious of the unknown. The idea that elephants had lived in the Klein Karoo before was not a concept that the Nature Conservation officials were happy with at all. Even senior officials were very reluctant to even contemplate giving us a permit to introduce a small herd. With all of the historical written evidence that they had lived in the area we still weren't getting anywhere. It felt like running up against a brick wall over and over again.

In my frustration, one morning at a staff meeting, I facetiously said to the team that I would hire a tame elephant and prove to them that they were wrong. My team all laughed but something got them going and we started discussing the idea seriously. After much debate and the feasibility of the idea had been thoroughly thought through, I approached John and Lindsay. They agreed with me and gave the go ahead to start the negotiations. I then approached the conservation officials with the idea and I must say that this started a wonderful relationship with them. They then understood we were serious about the task at hand and were prepared to do the research correctly. This had never been done before and little did we know that we would have the eyes of the world on us.

Interestingly, we had great support from many well-known elephant specialists who were very keen to see the results; but equally, we also had many elephant experts who said we were just wasting our time and that it was not a good idea at all. I suppose in hindsight it was rather revolutionary. My idea was to get hold of a tame elephant and bring it onto the reserve and have my team monitor every plant that it ate, the effect that its feet had on the vegetation, and if anything would germinate in the dung that it produced. The idea was that at the end of a certain period of time we would know what elephants would and would not eat, and if the animal had specific preferences which could create overgrazing and what would be the trampling effect of a four ton pachyderm on the Karoo vegetation. After lots of brainstorming we came up with a very workable plan – two of the students would have the job of observing the front end of the elephant, another would observe the trampling effect of both front and back feet and his task was also to collect a dung sample at random opportunities as and when they presented themselves. So, at all times there would be three researchers around the elephant monitoring its every move and looking for anything out of the ordinary.

I then approached the authorities with a very well laid out methodology and plan. It was really only at this stage that I began to think about where I was actually going to acquire an obliging elephant! We had made up the vegetation lists in such a way that it would help the two students with the upfront work as quickly as the elephant ate. The idea was that they would tick off the specific species that were consumed, what damage was done to the plant and how much of each plant was utilised.

After a rather interesting meeting, the conservation officials gave their permission and agreed to issue a permit to bring an elephant for a specific period of time. Only then did I start my search for a possible candidate. There were three possibilities and after numerous telephone calls I went on a reconnaissance trip. You can imagine the elephant owners' reactions to my request. They must have thought that I'd completely lost my marbles, arriving on their doorstep and asking to borrow an adult elephant for a while to do some research out in the middle of nowhere!

Fortunately, I met a man by the name of Ian Withers, and although it took some convincing, he soon understood the merits of what I was trying to do. He even came to the conclusion that this would be a great opportunity to assist in the much needed field of elephant research and conservation. Ian had an elephant bull that was about fourteen at the time. His name was Harry, a highly intelligent elephant that listened to commands but most importantly, was treated with love and respect by his trainers.

Just the logistics of moving an elephant is mind boggling, so training Harry to get in and out of the trailer started as soon as possible. Then short road trips were undertaken to get him used to the idea of going somewhere, but always returning to his home and friends. Ian and his team worked for months getting the enormous horse box for an elephant up to standard and of course getting Harry ready for the important work that he would do.

The day eventually arrived and Harry and helpers arrived on the reserve. My team was ready to get going on the task but had to be introduced to Harry properly and to learn the do's and don'ts around a tame elephant. Like: don't ever get between an elephant and a solid upright surface – they use every opportunity to rest their legs and lean over on anything solid, and if you just happen to be in the way, you'll have four tons of elephant bearing down on you which will crush the living daylights out of you. And the ellie, of course, won't even notice that he's hurt you!

My team was well prepared for the programme and we had selected areas close to the base so that Harry wouldn't have to move too far on his first day. The idea was to get him into as many different areas as possible, although there are only two main vegetation types on the reserve. Within those types are many different variations. In the Klein Karoo as with most vegetation types, the north-facing aspect gets more sun and the plants have adapted to this. I wanted to get Harry to sample as many as possible and to see what effect he would have on the vegetation.

The one aspect that I was very concerned about was the trampling effect. It was very important to me that the introduction of any species into the area would have a positive effect on the environment and, of course, when one looks at the size and shape of an elephant's foot you can't help but think that they would flatten everything they stood on. Having observed elephants from forests to grasslands and mixed bushveld, I was anxious to see what effect the movements of the elephant would have on the Karoo vegetation.

Off we went into the veld near the base; this area had been over-utilised to the extent that there was very little plant species diversity, so we didn't have too much of a problem noting what Harry ate. This did, however, give us the opportunity to observe very carefully the effect his big feet would have on the various plant species. It was quite remarkable to watch from less than a metre away how, when his foot landed on the ground, it would start at (for want of a better word) his heel and then curl its way over any object, being plant or rock and then finally land on the toe area. This cushioning effect allowed the plants to bend over to a point but not break, so when the foot was lifted again the plant would spring back into its normal position.

The team of Karlien and Rob at the front end were kept busy noting each plant and the amount taken from the plant and any damage that could be detected that could possibly kill the plant. It was an absolute revelation; Harry ate vegetation that none of the antelope species ate, and he utilised literally everything without doing the damage that is incorrectly associated with elephant behaviour.

During Harry's time with us we moved all around the reserve from dense acacia thickets, along the river lines to the most open, denuded north-facing slope that only had mesembryanthemums and crassula species on it. There was another aspect that occurred that I hadn't even contemplated; the press got word of what was going on and we were inundated with calls and requests for interviews.

The South African Broadcasting Corporation's television programme 50/50 asked to film the project. From this and other press releases we were to get an enormous amount of support from far and wide. It was an incredible experience to see an elephant in the Karoo, rather shocking at first, but after a while I realised that the environment was poorer for having 'lost' these magnificent creatures. I'm really glad that in a moment of frustration and anger I came up with this idea because it opened people's minds to greater understanding of elephants. And eventually it led to the introduction of a small herd to Sanbona, which, to this day is doing incredibly well, to the point that it would be difficult to visualise the reserve without their presence.

Just after Harry left the reserve I got a phone call from a local farmer asking me to come and identify a jaw bone that he had found on his farm. On arrival he showed me the lower half of a very large jaw bone that I knew immediately was from an elephant. I didn't confirm this with him, but asked if I could send it to the Museum for proper identification. This was done and it was confirmed, proving that they had existed in the area previously. There are now more than fifteen elephants on Sanbona and they are a natural part of the landscape.

CHAPTER 8

CHEETAH

We had cheetahs in the boma that were nearly ready for release into the main reserve. This was also going to be a first in the Klein Karoo. In South Africa space for predators is shrinking at an alarming rate. This brings with it the problem of predators moving onto formal agricultural areas and taking domestic livestock. They then become 'problem animals' and are vulnerable to being shot, trapped and sometimes even poisoned. Anne van Dyk has spent her whole life working with cheetahs. She has seen them become an endangered species and was one of the champions who worked so hard for many years to get the numbers out of the red and finally off the endangered list. Anne actually, at one stage, had ideas of moving to the Klein Karoo and carrying on her cheetah work and research near Montagu. She bought the farm Hoogfontein (meaning high spring). Hoogfontein farm was one of the very first that the Linton group bought to start the reserve. It is situated right in the middle of the 55,000 hectares, so it is a tribute to Anne that we were now going to be releasing cheetahs back into an area that had not seen them for centuries.

Cheetahs are really special animals to work with, and each and every one that I've worked with has had an individual character. We received two that had been stock thieves and trapped. A superb young male of about two years and a female of about three, one of the most delightful characters I've ever had the privilege of studying and working with. For monitoring purposes we put a tracking collar on the female and her tracking code on the telemetry set was 64; the male had an internal tracker inserted and his number was 00 so this is how they were named: Sixty Four and Zero Zero. We had created two cheetah bomas for our initial introduction. They are solitary cats and are hardly ever seen in pairs except when mating or when a coalition is formed, normally by brothers when they're old enough to leave their mother. So each of the cheetahs had its own enclosure that was close enough for them to see one another.

Something rather unusual happened when we opened the crates to let the first cheetah into the introduction boma. A pair of fish eagles flew over the area and started vocalising. The call of the African fish eagle is absolutely distinctive and, for me, is the bird call that epitomises the spirit of Africa. During the introduction I didn't see or hear the eagles calling; it was when I looked at the video afterwards that one of the students had taken, that we heard the call. I want to believe that it was nature's way of signalling her approval of what we were doing.

At every opportunity I would get down to the cheetahs and stand outside until everything was calm and then go and spend half an hour just moving slowly around the outside till Sixty Four and Zero Zero got used to me. They were in separate bomas, separated only by a fence. Both of them calmed down fairly quickly and relaxed in my presence; and after two weeks I started going into the boma itself with a camping chair and a book and I would read out loud for half an hour, letting Sixty Four and Zero Zero feel completely secure in the knowledge that I was not a threat. Sixty Four accepted me to the point that she would lie down about ten metres from the chair, and relaxed so much that even when I got up to go she would stay exactly where she lay.

I didn't have the time to sit in Zero Zero's boma as well, but he obviously took note of what was happening next door because when he was released, he was very relaxed around humans and many people have had the most incredible sightings of this big cheetah over the years. I was tracking him one afternoon when a strong signal told me he was very close. But I couldn't see him at all, so I walked closer to where the telemetry indicated and nearly stood on him. He was so well camouflaged that if he hadn't flicked his tail I wouldn't have seen him. The Karoo landscape looks open and bare but when an animal doesn't want to be seen they literally blend in with the environment.

Something else we did when feeding the cheetahs in the bomas was give sharp blasts on a referee's whistle when the vehicle approached the boma.. When feeding we would throw the meat over the fence and give another loud whistle. This was to get the cheetahs to associate food with a sound so that in the event of an escape we could entice them back with a whistle and some meat. This was to prove incredibly helpful some months later when another cheetah that was released broke a leg in a hunting accident. I was able to entice her close enough to dart just by using the whistle and a chunk of meat. We took her to the vet who did an X-ray and ascertained that it was a clean break so there was a good chance of recovery. He put the leg in a cast and we took her home. We put her in one of the smaller bomas and observed her. About two months later we decided to remove the cast to see if the bones had healed. At this time we had a very gung-ho student from Australia, a really nice young man who was keen to do everything, so when I asked for volunteers to help with the cheetah he was the first to come forward.

Due to the fact that the cheetah had been anaesthetised quite a few times and because she was very calm, I decided to attempt the removal of the cast without drugging her again. This turned out to be a poor judgement call on my part and when we had her strapped in a blanket and on her side I asked if her head was secure. The young Aussie said yes, so without physically checking I started removing the cast, at which point there was a rocking motion up near the head and what she did was chew through the blanket. A slight yelp erupted from the young student and I got a bit grumpy and told the team to hold on as I was nearly finished. As I finished and was giving the instructions on how to let her out of the blanket the young student turned snow white and said that he was going to faint. We released the cheetah and I went over to the student and asked what he was going on about. It was then that I saw that the tip of his finger had been removed. We gave first aid and one of the team took him straight off to hospital – the end of his finger had been cleanly severed by the cheetah. Luckily, the doctors did a great repair job. You can just imagine the stories that are going to be told around Australia of how he lost his fingertip to a cheetah. Fortunately we had all just done an intensive week's first aid course where we had learnt most of the basics and even how to deliver a baby, so I'm happy to say that the team handled this medical crisis with great aplomb.

When Sixty Four and Zero Zero were released into the reserve we had quite a to-do, with lots of press and the owners arriving with VIPs to witness the release. If you ask any person who has done this kind of thing what is the best method, they'll all tell you: no noise or people around, just open the gate quietly and walk away and let the animals decide when they want to walk out. This was not going to happen. Understandably, the press wanted pictures of the cheetah walking out of the closed space and into freedom. This would be the first time that cheetahs would walk freely in the Karoo for at least 250 years. To do this we put a springbok carcass about fifteen metres outside the gate and staked it to a peg, all the while showing it to the cheetahs so as to whet their appetites.

From a previous release we had learned the hard way. What had happened then was that I hadn't pegged the carcass down and the animal walked out of the boma, grabbed the carcass and dragged it straight back inside.

When the guests arrived we arranged the vehicles at 90 degrees to the gate so as to get the best pictures but also not to stress the animals. One of the press folk asked if he could put a small video camera near the carcass so as to get the cheetah coming out of the boma head on. It was explained that he could do this with pleasure but if anything happened to the camera he couldn't hold us responsible in any way. With everything ready, Ryno and Paul opened the gate and walked away. Within about a minute Zero Zero walked out and started eating on the carcass, bringing a big sigh of relief from the wildlife team. Just when everything was relaxed and calm and the only noise you could hear was the clicking of cameras, Zero Zero jumped into the air and stamped his two front paws repeatedly onto a small pile of rocks. Dust flew in all directions, stones flying but as quickly as it had started he stopped and went back to eating.

The morning was a great success and the Klein Karoo had cheetahs roaming the veld for the first time in a couple of centuries. After the cheetahs moved off we went to collect the video camera and it was then that we realised what Zero Zero had stomped on. The video camera was completely destroyed, bent and buckled, but the owner still had a smile on his face. When I asked why he pointed at the tape which was probably the only part of the camera that looked in good shape. That footage still gets aired on various TV channels to this day. I suspect that it must have been the whirring sound of the video camera that aggravated Zero Zero to the point of aggression.

A very special event occurred on the reserve about six months later. Sixty Four gave birth to four cubs, high up in the hills of Hoogfontein. It was fitting that the first cheetah cubs born in the Klein Karoo for centuries would be born on Anne Van Dyk's old farm. Sixty Four was an incredibly good provider and her cubs did very well but what was remarkable was that all of the work in the boma had paid off. She remained very relaxed around humans and the rangers could take guests on foot to within ten metres of a wild cheetah with cubs, an experience that would live with them forever.

We know the cheetah as *Acinonyx jubatus* which reflects its uniqueness; it's inspired by its most characteristic features and traits that it shares with no other member of the cat family. *Acinonyx* is thought to derive from two ancient Greek words – *akantha* which means thorn, and *onux* which means claw, this being a reference to the cheetah's dog-like claws. There is an alternative to this in the words *A* and *Kino*, Greek for 'not move' which is a description of the cheetah's poorly developed ability to retract its claws. The species name, *jubatus*, is derived from Latin *iabatus* which means maned. This refers to the grey-coloured mane that the cheetah young are born with.

The king cheetah has a very interesting coat pattern. It is so different that at one stage it was thought to be a leopard-cross-cheetah. Back in 1927 when it was first described it was thought to be an entirely new species and called *Acinonyx rex*. The difference is only skin deep; the pattern arises from a recessive mutation to a single gene that controls the formation of spots. This makes the king cheetah the cheetah version of a tabby cat.

Partial albino cheetahs have been recorded and the cheetah in the Sahara are very pale, bordering on being white with very pale spots. Way back in 1608 a white cheetah was presented to the Mughal Emperor Jahangir, being described as white with bluish spots.

CHAPTER 9

THE CUBS

Jabulani and Queen were getting along just fine and their condition was excellent. At feeding times Jabulani would, as per norm, get in first and gobble down quickly as much as possible. However, we always made sure that there was enough for Queen to fill up on. At this stage they were being fed whole carcasses so that they would know how to handle them when the cubs arrived. Up to this point in their lives they had only ever been fed on large chunks of meat and it was rather funny watching Jabulani fumbling around trying to work out where to start on a whole carcass. It was vital that they learnt how to deal with a carcass as when their cubs arrived they would receive only natural prey from Queen. Everything that the cubs experienced and learnt from their parents would be vital to their future as wild lions.

The suspected date of arrival of the cubs was drawing near. The gestation period of a lion is on average 110 days, but birthing can occur anywhere between 105 and 113 days. It was now time to separate Queen from Jabulani as this is what would happen in the wild. As explained earlier, under natural conditions a lioness will separate from the pride and find a thicket or a cave to give birth in, away from the male. Separating them would also give us the opportunity to do a scan on Queen to confirm her status.

Lions don't show their pregnancy literally till the last week before giving birth. The date was set and Johan came down from Shamwari and Professor Henk Bertschinger arrived with his equipment to do the scan on Queen. We darted the two and took Jabulani straight to his new boma at Brakrivier and Queen to be scanned. Henk started the scan, and there wasn't a peep out of anybody. You could hear a pin drop. The scan is done rectally and the images show up on a screen, which is a green colour and quite misty so if you don't know what to look for, it can be decidedly frustrating. After a while, Henk pointed to the screen and outlined three cubs. Cheers erupted; it was as if we were all expectant parents and excited speculation started. One of the students even asked him if he could sex the cubs – which he couldn't.

We took Queen to her new boma to come around and the anticipation of the arrival of the cubs was tangible. With all of the other goings-on, the reserve was a busy place for the wildlife team. The pressure was quite intense. For all of the introductions of the different species we had to give motivations. So over and above the normal day-to-day work we were bombarded with endless requests for paperwork that, in retrospect, was so absolutely unnecessary. The amount of paperwork required by the authorities is mind-numbing – there must be somebody there with nothing to do all day except work out how to get more superfluous information from wildlife managers to cover every aspect, to the point of being tautological.

To make sure that I could answer any questions regarding Jabulani and Queen I went down to Brakrivier every morning and evening to check on them. I must say that the drive down there before dawn was very special indeed and it's those moments that remind you forcibly why you chose to be a conservationist. On arrival at the boma I would drive down the western sector and make sure that I could see Queen and check her condition. If she was out in the open I would know the cubs had not been born yet and move on to Jabulani and check on him.

At sunrise one morning I drove to Queen's section of the boma and there she was sitting out in the open. I noted that she had a smudge on her side but other than that she looked pretty good. As I started to move on to Jabulani a movement caught my eye. In a small dip right in front of her something had moved. I stopped, not even thinking it could be a cub as like all lions she would have had them in the thicket. I waited for a while and she leant forward and started licking something in the dip. Right there, out in the open she licked a cub right out of the dip towards her belly. I couldn't believe what I was seeing – a perfect, snow-white cub out in the open for all the world to see! Something was wrong; lions just don't do this. I wanted to spread the wonderful news to the team but because of her behaviour I was very reluctant and decided to sit there and just watch what was going on. Of course, all of the negative things that could happen were going through my mind. My worst nightmare was that if she didn't rear the cubs herself and we had to hand-rear them it would stop the project in its tracks right there and then.

Sitting there looking into the boma at Queen through a pair of binoculars I was jolted out of these negative thoughts by seeing a second cub being licked and enticed towards her teats, where the first one was busy at the milk bar.

This behaviour was totally foreign and out of place. I couldn't understand why she wasn't in the thicket. Realising that all three cubs had been born, seeing them all suckling and that everything appeared in order, I drove away to tell the team the news. Phone calls to the bosses, Adrian and Gaston, were very well received but I was still very worried that she was out in the open. During the day Ryno, Paul and Karlien went to check on the new arrivals and both came back and said the same thing – three cubs all healthy and suckling from Queen but still out in the open. At the end of the day I drove down and nothing had changed. I sat for about half an hour with my spotting scope examining every detail that I could and my observations told me that the youngsters were in good health, Queen was in good health – she was just not acting like a normal wild lion would.

That night was a mixture of elation that the cubs had been born and were healthy, mixed with a terrible worry that Queen didn't know what she was doing; yet she appeared to be a very good mother.

Every day was the same for the next seven days – I visited twice a day and the team checked on her and fed both Jabulani and Queen but there was no change. She and her cubs were out in the open, exposed to the elements.

Day eight arrived and, as usual, I was down at the boma as the sun crested the Karoo koppies and Queen was nowhere to be seen. I drove around the enclosure three times without a glimpse of her and there were no indications that she had escaped. No lion spoor outside or any other indication that she had got out.

Against my own rules I walked around the outside section where my vehicle couldn't go. I was so anxious to find her that I spent most of the morning searching for her. Nothing at all, I couldn't see or hear anything, so I decided to climb up the feeding platform and observe. For what felt like hours I sat on the feeding platform, staring into the thicket and any area that I thought could hide a lion. I couldn't see anything at all that vaguely resembled a lion. Then suddenly … a small movement and then another and all of a sudden right there in front of me was Queen. She had moved into the thicket that we'd incorporated into the boma for her to give birth in and that she hadn't utilised at all.

For the next four weeks we had completely the opposite problem. We hardly got any sightings of the cubs. I did some reading and found out that in the lion breeding centres they remove the cubs on the fifth day after birth. Knowing that Queen had produced litters in the breeding centre, she was probably used to the fact that she was going to lose the cubs, and when she realised that we weren't going to take them away from her, she became a normal lion and hid them.

Now I had a very different problem. Everybody wanted photographs and to know how the cubs were and what sex they were. This was literally impossible without going into the boma itself. There were some days that I wouldn't even see Queen let alone a cub, so the reports that I would send through didn't make for exciting reading and this went on for another four weeks. Then one morning, as I approached the boma, on top of a small dune, I could see Queen. She looked healthy and her attitude towards the vehicle was exactly what we wanted to see. I was having a good look at her through the binoculars when a cub jumped over her back and landed in a heap, followed quickly by the other two who then proceeded to have a rough and tumble right out in the open. This is what I'd been expecting to happen but nothing really prepares one for a moment like this! Then it was a quick rummage around to get the camera out and take the first pictures for the bosses in four weeks. The cubs looked healthy and I could sex two of them but was unsure of the third, so in my paperwork for about a week it was two males and one question mark.

The word had got out that the cubs were in the open and Queen was very relaxed so we decided to let a small number of visitors view the cubs from game viewing vehicles. We briefed the rangers on the do's and don'ts and it must be said that they were a really responsible group of men and women who understood the sensitivity of the programme and became a great help in the monitoring process. The rangers got very good reactions from their guests and whenever I had the opportunity to speak to any of them it was very interesting to hear their comments. I remember having to do many radio and television interviews – it was a very busy time.

One rather interesting aspect about TV crews is that the camera operator always wants to get closer, even with the powerful lenses they have. One newsman wanted to interview me with a lion in the background so I asked the team to find the reserve pride and give me an update on what they'd been doing. Luckily the male was quite close to the office so we drove down to find him in the shade of an acacia tree. He was about 150 metres away and very laid-back so I set everybody up and gave them the briefing – no fast movements and to do exactly what I say at all times!

The lion had a distended stomach so I knew that he'd eaten recently and it was in the middle of the day so he could possibly even fall asleep. However, one must never make the mistake of becoming complacent around any wild animal. So with the interviewer and me facing the lion the cameraman started moving around us and I asked him not to get between me and the lion as I needed to keep an eye on him at all times. He then told the two of us to turn around so that the lion could be in the picture behind us. It was then that I explained to him that the big, fat pussy cat relaxed under the tree could travel 100 metres in about three seconds and be upon us in five. There was a stunned silence and no further requests to get closer or for me to turn my back on the lion. When lions are inactive they look relaxed and peace-loving; however, one wrong move and they react instantaneously to a threat. From being asleep to up and ready takes a millisecond.

It was an incredibly busy time and over and above our normal work we got calls from all over the world with people requesting photographs and interviews on a daily basis. Ryno's home language is Afrikaans so I asked him to handle the requests for interviews in that language. He coped very well, as it can be a nerve-racking experience knowing that your interview is live to the public.

When Queen's cubs were six weeks old we let Jabulani into the centre boma. He and Queen rubbed necks and heads through the wire and Jabulani then lay in the corner of the boma right next to the cubs. It didn't take long for the cubs to be playing right beside him, separated by the fence, and for the next two weeks every time I went to the boma the three cubs would be playing in the corner right next to Jabulani.

It was time to let them be a family unit and put them together. Of course this is a very dangerous moment in the lives of the cubs as anything can happen. However, judging by the actions so far there was a good chance that everything would be okay. The day was set and very quietly the gate was opened, with the dart gun on hand in case the worst case scenario occurred.

Against all expectations, Jabulani rushed into the boma growling and roaring. I was absolutely horrified; I couldn't believe what I was seeing. Queen manoeuvred herself between him and the cubs but he kept on trying to get at them. With the cubs running around and the big lion roaring, it was extremely frightening.

The cubs ran up to the top of a small dune and Queen positioned herself just in front of them. Jabulani then made a rush at the cubs – I don't think I've ever seen anything so incredibly fast and decisive. Queen charged Jabulani and because he was halfway up the dune and side on to her he was slightly off balance when she hit him full force in the middle of his chest. The oomph on impact was substantial and you could hear the air knocked out of his lungs as he somersaulted down the dune, crashlanding in the most undignified heap. He just lay there, all bravado gone.

Queen then brought the cubs to him and from that moment on he became the most incredible father. It was an amazing sight to see, this great big lion smacked so hard that he was cartwheeled off the ground and winded to the point of submission. Queen had become a formidable mother,

her whole demeanour had changed since she arrived on the reserve and she would often give a snarl by just lifting one side of her lip in warning if she felt anything threatened her family.

The feeding was done in such a way that the cubs never saw where Queen and Jabulani brought it from. This was vital as we couldn't allow the cubs to associate food with people or vehicles. Now that we knew all of the cubs were male, the options were narrowed down and we started the preparation for the next aspect of the programme.

CHAPTER 10

THE FLOOD

Sanbona is in an arid zone with an average of 400 mm of rain in the southern area and 250 mm in the north. When the Bellair Dam in the north is full, the earth wall pushes back the water for about five kilometres making it a substantial body of water, ideal for waterfowl, fish, hippo and other aquatic species such as terrapins and otters.

I was having a weekend getaway with Simmy in Knysna when I got a phone call to say that a Welsh camera crew was arriving at the reserve and that I had to cut my weekend short and get back to the reserve to show them around and do some interviews and generally entertain them. I told the team that I would arrive back on the reserve that Sunday afternoon, take them out for a drive and later have dinner with them. It was beautiful weather, not hot but very pleasant. We did a game drive and I must say that the Welsh folk were wonderful and it was my first real taste of the Welsh language. We would stop and look at something, the lady interpreter would ask me a question in English, I would answer and she would then repeat my answer to the camera in Welsh.

To an English-speaking person who speaks Afrikaans and has a working knowledge of Zulu plus a meagre understanding of French, Welsh is unlike anything I've ever heard. A beautiful, lilting, sing-song style language that doesn't give one any hint of what's being said, except, of course, to other Welsh speakers. It intrigued me. The crew must have thought I was crazy because I kept asking the lady to repeat what she'd said.

That evening at dinner one of the lodge staff told me that it had started raining. It was 9 pm and because just two hours earlier there had been no indication whatsoever that it would rain I went outside to check, as I thought it might be a joke. Sure enough, it was a beautiful steady rain but the temperature had plummeted and we all scrambled for jerseys and jackets. At the end of the evening I arranged to collect the Welsh crew first thing in the morning so that we could get them the footage they still wanted. On the drive back to the house I was really happy as it was only 22 March and we were getting a very early rain. The reserve falls into the winter rainfall area and the summer had been dry with no rain for quite a while. I was woken at about two o'clock in the morning by the sound of driving rain – this was not normal but still very welcome. I went back to sleep and at the normal early hour of four o'clock I was in my office doing the correspondence and by six o'clock it hadn't stopped raining.

Instead of having our normal morning meeting I radioed Ryno and asked him and his team to inspect the southern side of the dam and the roads on their way to the office. The radio soon crackled into life that the causeway on the western side of the dam was a metre under water. This is normally a dry, dusty stretch of about 400 metres in width so, on hearing this I jumped into a vehicle and drove up there. We have had sporadic flooding incidents but nothing like this. I couldn't believe my eyes – a raging torrent with whole acacia trees and all sorts of debris flowing rapidly into the dam. I could just make out Ryno and the team on the far side through the torrential rain. I radioed Theresa to phone our neighbours in the catchment area and to apologise to the Welsh media contingent that I wouldn't be joining them for breakfast. The rainfall figures were disturbing. In the southern catchment area 180 mm had fallen overnight, and to the west our neighbours had recorded 160 mm in five hours and it was still pouring down. I instructed the staff to be extremely cautious and only to drive if absolutely necessary. By 11 am the neighbours had recorded over 300 mm in less than 24 hours, this being more than the yearly average!

At 3 pm the dam's outlet started to overflow; the dam was completed in 1919 and the outlet had not once overflowed. It soon became apparent that the area around the staff housing and workshop, which was situated at the base of the dam wall, was in danger of flooding and Johan, the maintenance manager, whose farm the company had bought, moved all of the tractors and vehicles to higher ground. I instructed the staff to get all of their belongings and pack them on the vehicles under plastic to protect them from the rain and to move to the southern side of the dam and that nobody was to be below the dam. I was in radio contact with the police and most of the neighbours along the riverine area below the dam that was starting to flood. I gave a warning that they should get all people and livestock to higher ground.

I then asked Johan to open the irrigation pipe that was underneath the dam wall and when he sarcastically laughed at me, I told him to take Theresa, who was his daughter and my personal assistant, to safety. Through the afternoon Ryno and Karlien kept me up-to-date as to what was happening in the south and Kevin and Karen (at the lodge) kept the guests happy. At about four o'clock Alison, a ranger who lived in the same house as I did, arrived and I told her to get all her belongings packed up and Simmy tried to get as many of our irreplaceable things into the back of her Renault as possible.

The idea was that I was going to move south of the dam (which is above the dam wall's height) for the night, as I'd sent everyone else to safety around the reserve. By the time I'd made sure everything was as safe as it could be, I drove over the ridge. It was an amazing sight to see – the water had backed up so far I couldn't get through to the southern area, leaving me no choice but to remain in the north and patrol it. On top of the ridge, which is called Boesmansrand (Bushman's ridge), I noticed, to my amazement, terrapins climbing the ridges. Seeing this sent alarm bells through my whole being – why would a terrapin climb a small mountain other than for survival?

At about 6.30 pm it suddenly began to get dark and it was still raining steadily. I decided to inspect the dam wall one last time and then head off to safety. At the base of the dam wall Simmy and I started to climb up the side to inspect the top when, all of a sudden, a gust of wind brought a wave of water straight over the front. I shouted at her to get back to the vehicle and we scrambled down as fast as we could. Just before I drove off I looked up and saw a strange sight. There was a flock of sacred ibises on top of the wall, and as each wave broke over the wall they would hop into the air and then land again.

While driving away I listened on the two-way radio system and heard how the police helicopter had rescued a farmer and his staff from an island in the centre of the river. I informed the police to evacuate anybody in close proximity to the river as I thought that the dam wall could possibly collapse. After warning as many people as possible, I told the team that I was going to take Simmy and Alison to the local hot water springs as I knew they had accommodation and would put us up for the night. We were cold, wet and miserable, with no electricity or food. During the day I had phoned Adrian and Gaston and told them what was happening, so everybody was up to speed if we needed extra assistance for whatever reason.

At five the next morning we started making our way back to the reserve. The road was completely washed away in sections and I had to negotiate our way over erosion gullies and around debris. When we rounded the corner where you could see the riverine area for the first time, it took a while to comprehend what we were looking at in the low light. The devastation was surreal – savage annihilation, totally unexpected. Where, just yesterday, the riverine area had been dense with thickets of acacia and rhus trees, there was nothing, just a sheet of silt-laden water stretching up to a kilometre wide in some places.

I reached my house and the damage was terrible – we'd lost everything that wasn't in the car with us. I walked around to the office and as I stood looking at the devastation

I couldn't quite work it out ... there was something missing. It then dawned on me that I couldn't see the workshop! It was just gone, a brand new workshop facility had just disappeared. It had been a big building, about 30 by 20 metres and it had just been swept away. I then ran down to the staff quarters and all four houses were flattened. Just pieces of foundation and brick work were sticking out of the mud.

These were brand new homes that the staff had just moved into. The water tank for the house and workshop had stood on a 10 metre high stand and when the dam wall gave in, the force of the water knocked it over and the 5,000 litre water tank fell straight through the roof and landed on top of our double bed, smashing it beyond recognition. It was just as well that I'd made the decision not to stay as Simmy and I wouldn't have survived had we slept at the house. We had lost everything in the house – sadly, a lot of history went down the river as thousands of slides which I was converting to digital jpegs had just disappeared. When Simmy was going through what remained of the bedroom, she told me that my three pairs of hiking boots were now just three right foot boots. The lefts had disappeared, along with all of my clothes and, sad to say, a really rather decent selection of red wines and some fine whiskies.

I radioed the team to check that all of the staff were accounted for and safe. Fortunately, they had stayed together and all were safe except that one of the fencing teams and some of the maintenance staff were missing. I was driving towards the student centre to co-ordinate a clean-up and on the way a red and white Air Force Oryx helicopter landed near us. They asked if we were safe and did we have water.

Over the noise of the rotor wash I told them that I was looking for a fencing team and some staff members. They offered to fly me to the areas that I couldn't reach by road. We found the fencing team safe in an old homestead away from the river, and the maintenance staff members were thankfully also safe.

They then flew a circuit down the river line below the dam wall so that I could ascertain the damage. It was too terrible. I saw a dead eland stuck in a tree, one of many casualties that we found. All of the fences in the river line and most of the acacia thickets had disappeared. The helicopter crew dropped me off and I can't tell you what a relief it was to know that all of the staff were safe. The South African Air Force has some of the most incredibly brave people. Without any hesitation they sprang into action, putting the public's lives before their own. I was very fortunate to meet up with the crew at a later date and had the opportunity of thanking them for their selfless commitment.

I have subsequently made a documentary film on how the same squadron put their lives on the line for a group of Russian scientists when their ship got stuck in the Antarctic ice. The Air Force must be commended for the rescue work they so selflessly do. What I didn't know at the time was that there was a reporter in the helicopter and he took a photo of me looking out at the devastation. My face said it all, and that picture went around the world and was on the front page of the Cape Argus.

One of the worst telephone calls I've ever had to make I did that day – it's very difficult to explain to your bosses that one of the main features of the reserve was no longer there. Over the next five days I had to drive to the southern area to get cellular reception as the telephone lines were also down. I did my best to keep Adrian and Gaston up-to-date.

Out of a disaster like this come some rather strange results. Firstly the dust! I can't express how bad the dust was; it came from the silt, finer than talcum powder which got in everywhere. The overhead power lines leading to the reserve had been washed away and it took five days before we had electricity again. Without it you can't pump water, so no water for five days. I don't have to tell you what it's like to go without flowing taps for five days. We banned all swimming in the pool and used buckets to collect the water and boil it for washing and cooking purposes. The fridges and deep freezes were off so there was another problem to contend with.

The next couple of weeks were spent cleaning up and reconstructing fences and staff homes so that the reserve could start functioning normally as quickly as possible. The government officials and the Mayor of Montagu arrived and surveyed the damage and promised to do what they could. All of the people affected by the floods in the town of Montagu were to receive a small pay-out, so I submitted all of the paperwork for our junior staff to benefit from this as a lot of them had lost many of their precious possessions. For three years I sent them reminders and am very sad to say that to this day those politicians' promises were never fulfilled. Whenever I see those members of staff, I get a pointed reminder about the deceitful way they promised the team that they would get some form of compensation. I would love to name and shame them but the law is an ass and instead of the officials getting even a rap on the knuckles I would probably be sent to jail.

When Simmy found my three right-footed boots, in what was left of my cupboard, we both knew instantly what to do with them. In Wynberg at the top of a steep hill is a traffic light where a man begs for money from passing cars. He only has one leg and we both immediately thought of him, so when we went down to Cape Town we went to drop them off with him. You can imagine our disappointment when we saw – Murphy's Law – that it was the right leg that he had lost.

CHAPTER 11

More Lions and a Great Man

The reserve was visited by Linda Tucker, who was trying to get three white lions from the zoo in Johannesburg. She asked for our support at a meeting that was to be held with the different parties involved with the white lions at the zoo. This was to be held in Johannesburg so I went up and before the meeting, had lunch with a man who has influenced me greatly and is a wonderful mentor, Dr Ian Player. Dr Player has been an inspiration to generations of conservationists, and has over decades of service to conservation, achieved a great deal. But the one accomplishment that he's most well-known for was his and his team's involvement in saving the white rhino. It is doubtful if there would be any rhinos in the wild today if it weren't for their efforts. I had the honour of introducing Ian to the Klein Karoo, an environment that he hadn't been in before. His reaction to the area was very interesting; he's an extremely spiritual person and the Karoo has great landscapes and silence that create an atmosphere for introspection. I know he enjoyed his time there.

Ian gave me some sage advice that he'd received as a young man: "Develop the thickness of skin of the animals that you love so much as you'll need it to defend yourself, not from the animals but from humans."

Somebody had the idea of hiring a donkey cart in which to show Ian around. Kevin found a man close to the reserve who had a cart, and on the due day he arrived with possibly the world's wonkiest cart. Ian climbed on board and away they went. Without knowing it he had a near-death experience because when the cart came back the left side wheel was in the process of coming apart. No sooner had he climbed out, than the cart moved off and the wheel rolled away into the blue yonder. I'm so glad we didn't cause the demise of this great man that day!

The meeting was quite a success and I'm sure Ian's role had a lot to do with it. My part was to assure the people that, if necessary, we had the facilities and the know-how to look after the lions until Linda found a permanent place for them. More work for the wildlife team, as we would have our own programme to work to as well as three extra lions to look after in the boma. I then had to apply for a permit from the conservation authorities who we had just assured that we were not starting a white lion breeding facility. You can imagine how much explaining I had to do to get around that one... We eventually got the permit and the lions arrived by air.

The airstrip on the reserve is not tarred but the surface is even and in good condition so that a reasonably large twin-engine aircraft can land safely. We did have one major problem though, as every time we put up a windsock the elephants would rip it down and destroy it. Even putting upturned sharp rocks all the way around the pole hadn't deterred them, so when an aircraft was expected somebody had to go out and put up the windsock and take it down when the aircraft left.

We put the three lions into the boma that Jabulani had stayed in while Queen gave birth. So there was still a full boma between the two families. For a few days there was a bit of vocalisation between the two but it soon settled down to being more natural behaviour. The adult female was a very fine lion, very solid and even though she had been in captivity her whole life she had good musculature. The two youngsters, one male and one female were in that typical six- to ten-month gangly stage but in good health and displayed a pleasant attitude, staying away from the fence when approached.

The team and I accepted them as though they were our own and I checked on those lions every day that they were on the reserve. They were a pleasure to work with and observe but something really strange happened on a couple of occasions. When inspecting the lions I found polished pink rose quartz semi-precious stones scattered around inside their boma. I asked the team about this and they assured me that it was not they who were doing some kind of strange ritual. To this day nobody has come forward and told me exactly who or what scattered them there, and of course I wanted to know why. Over and above my curiosity it was a damned nuisance because I would have to go and gather them up in case the lions swallowed them and choked.

Young lions are not the most intelligent creatures and you never know what they'll do with a shiny new foreign object in their living area. I know their digestive system is quite hardy but one never knows what a stone like that would do to their alimentary canal. It only happened a few times but it was a great worry to me because we never ascertained which irresponsible person had thrown those stones. I still get the shivers just thinking about how I would have had to explain if one of them had become ill after ingesting a semi-precious stone.

The boma that housed the three was close to our new pride and it had a section of fence against the open area where we were going to release Jabulani, Queen and the cubs. Previous experience had taught us that we would have to create an extra fence offset from the boma to prevent any aggression. We'd been preparing for so long and this was another aspect that could create problems in the future, as we now had three extra lions to distract the pride from a normal life. And we would still be feeding these three till a new home was found for them.

CHAPTER 12

Biology Lesson Lions 101

Lions are the only distinctly sociable cats, their social structure having been studied in depth for many years which means there are many great works and papers for conservationists to refer to. They live and hunt in prides that can vary in number from a few individuals up to 30. The average in the Kruger National Park is 12. Lions hunt in areas where game is concentrated and the cover is sparse so it is to their advantage to hunt collectively. They are capable of living in a wide range of habitats – they go deep into desert areas where there are river lines so they do very well in areas like the Kalahari and the Sahel in west Africa. The only habitat that they don't appear in is the forest areas of central Africa. They will drink water when it is available but they are not water dependent as they can get their moisture requirements from their prey. The most important requirement for the survival of lions is enough food in the form of medium to large-sized game.

They don't need dense vegetation to hunt in and the sparsest desert scrub will fulfil their hunting necessities, but they do require shade to lie down in and protect them from the heat of the day. I have seen a line of lions sharing the shade cast by the wing of a single engine Cessna Caravan aeroplane. Just by the way, with reference to that, I've noticed that lions have a thing about tyres. I've seen them chewing both aircraft and vehicle tyres which has resulted in quite a few delays and some rather irate pilots. Lions are predominantly night hunters and their vision has adapted to this. They have round pupils and their eyes are proportionally larger than animals of comparable size. Contrary to popular belief a lion's eyes do not glow in the dark; however, they have a special reflective coating that is so sensitive it will even reflect starlight. This coating ensures that every available photon of light is directed to the cells of the retina. If you look under a lion's eye you'll see a lighter band of hair. This also reflects any available light into the eye. Lions also have a nictitating membrane; *nictare* is the Latin word for 'to blink.' It is a third eyelid that can be drawn across the eye to protect it but it also moistens the eye while the lion maintains visibility. This membrane is often seen in the eyes of birds of prey. Lions' eyes have limited side to side movement and this is why they move their head to look in a specific direction.

Lions have a well-developed sense of smell, marking their territories by depositing scent, normally in the form of urinating onto bushes, trees and rocks. Lions have two distinct behavioural patterns when urinating. The first is a normal urination when the animal squats close to the ground and releases a stream of urine, sometimes accompanied by scraping. The second is spray marking, when one or more sprays of urine are ejected by the animal while it stands with its hindquarters towards the object that it wants to mark, with the tail lifted vertically. The fluid deposited in both of these behaviours comprises normal urine mixed with a whitish lipid precipitate which is a hardener. When deposited as normal urine the lipids do not leave distinct marks on the ground but when sprayed on vertical objects the lipids will stick to the surface and, in time, turn into a black tar-like substance which marks the object for a long time and almost becomes weatherproof.

This effect is intensified as the lions tend to spray-mark the same objects on their territorial patrols. They also leave scratch marks on trees, the males standing on their hind legs and stretching their forelegs as high along the bark of a tree as they can. They then hook their claws into the bark and pull down, creating long scratches along the trunk of the tree. Because lions have scent glands along their claws, this is both an impressive visual sign and a scent mark. The scratch marks can be so long and deeply gouged into the wood that they scar the tree for the rest of its life.

You might have seen pictures of lions grimacing or pulling a strange face almost like a snarl. This is called Flehmen – it's a German word which literally means bending the upper lip. It is used when examining scents left by other animals either of the same species or of their prey. The action helps to expose the vomero nasal organ and draws scent molecules back toward the organ and this allows the lion to determine certain factors, including a female's state of oestrus and even how long ago an animal passed by. The vomero nasal organ (also known as the Jacobson's gland) is a chemoreceptor organ. It is named for its closeness to the vomer and nasal bones and is particularly well developed in lions. Flehmen is not exclusive to lions – giraffes, horses and even domestic cats use it to establish what messages have been left by other animals.

Lions have good but not exceptional hearing. I have had the opportunity of testing this on a few occasions and I must say that I was rather surprised. A pride had fallen asleep once around the vehicle and I needed to move off. I know from experience that lions that are startled go into protective mode and I didn't want that, so I spoke a few words out of the window. Nothing happened, so I tried again a bit louder. Nothing, then I did what I always do around elephants. I said loudly 'starting'. This got their attention and I could then start the vehicle without creating too much alarm. The rounded ears can be swivelled through a wide arc which then directs the sound to the ear hole, so they can hear sounds from a distance and have directional accuracy as well. The backs of the ears are black with a white dot in the middle.

The mouth of a lion holds the secret to its identity as along the top lip are a row of vibrissae, which to us mere mortals are known as whiskers. They don't have any nerves in them but they do have very special sensory cells. The whiskers are different from other hairs because they are implanted in a follicle that is sealed in a blood capsule, called a blood sinus. When the whisker touches something and bends, the blood in the sinus is pushed to one side or the other, the blood then amplifying the movement and the receptors at the base of the whiskers detect miniscule deflections. In lions the follicles are surrounded by a sophisticated sheath of muscle tissue which is used to move the whiskers. They are so sensitive that they can even detect air movement and a great deal of time is spent making sure that their whiskers are clean.

The longest whiskers are on its upper lip, called the mystacial whiskers. Those above the eyes are called the supercilliary whiskers and those on either cheek are the genal whiskers. Whiskers are not unique to the head; they are found on the paws as well, known as carpal hairs. They are used to feel vibrations in the ground. The arrangement of whiskers is not random as a good picture will show an ordered grid of arcs and rows of whiskers, shorter in front and longer at the rear. These are like lion fingerprints and the arrangement from the nose backwards is unique to each lion. For research purposes this is very handy; however, with so few lions on the reserve we didn't have to rely on the whisker identification method as each lion had other unique features and all had different personalities. Every whisker emanates from a dark spot on the muzzle. These whisker spots are aligned in rows, and each set of rows is unique. However, the rows on both sides of the face differ, so to positively identify a lion you have to be able to see both left and right rows. Once you can identify both patterns of whiskers there is very little chance of mistaking any lion for another.

Unlike any other cat, the lion possesses a tuft of hair at the end of its tail. This tuft of dark-coloured hair hides a spur of bone which is separate from the other bones in the tail. Its function is not known, and it is not present in all lions. In white lions this tuft is white. I was once told by a spiritualist that in certain lions there is a hook hidden in the tuft, especially in the white lions and the lion uses this hook in the spirit world to gather people together. At every opportunity I have inspected the tails for this hook but so far I haven't seen or felt this anomaly. Females also use their raised tail as a 'follow me' signal for the cubs. They also use it to signal to each other during a hunt.

The lion's teeth are well adapted for killing and eating their prey. The large canine teeth are spaced so that they can slip between the cervical vertebrae of some of their prey and sever the spinal cord. The back teeth, called carnassials instead of molars, are shaped in such a way that it makes them work like a pair of scissors. The lion's jaw is not capable of moving side-to-side like ours. This helps keep the carnassial teeth in alignment for cutting. The rest of the teeth are conical, and designed for cutting and tearing. Lions don't chew their food, they swallow it in large chunks as quickly as possible. Lions only use one side of their mouth at a time because of the inability of the jaw to move from side to side.

The tongue is covered with rough spines, called papillae. These help the lion to scrape meat off bones. Whenever we used to dart a lion we would allow the students to feel the lion's rough raspy tongue. I'm sure that some of the old stories about lions actually licking away a man's feet are quite plausible. The fact that they groom themselves and each other without damaging their fur is testament to just how tough a lion's fur and skin are.

The digestive system of the lion is very similar to that of humans. Their diet consists of meat which is fairly easy to digest. The digestive systems of herbivores are rather more complex in comparison as they need elaborate digestive mechanisms to break down cellulose. Lions, and in fact all cats, have the shortest digestive tracts of all animals. A lion's body temperature ranges from 38.05 to 39.16°C.

No description of lion biology would be complete without mention of the roar. Only four cats can roar: the lion, tiger, leopard and jaguar. They are assigned to the genus *Panthera* because they can roar. Of these four cats, the lion roars the most. It was always understood that roaring was made possible by a two-piece hyoid bone in the throat. However, there is research that suggests that this is not the case and that the larynx which, in lions, is longer than any of the other cats, is the more important factor. It is possible that this is why the lion's roar is the loudest of the four members of the *Panthera* family.

Adrian and I were once up in the Kalahari in the middle of summer and after a long hot day of raptor monitoring we made camp. The temperature was nearly 50°C so we decided to sleep on a mattress on the ground under the stars instead of putting up a tent. At about four in the morning I heard lions roaring in the distance. Adrian woke up and I assured him that the lions were a good few kilometres away. About ten minutes later Adrian shook me awake and said, "Listen Dad, there's a lion right here!" I could still hear the lions roaring in the distance and tried to pacify him, but what I hadn't heard were the footfalls of a lion padding down the road. This was another lion, a massive male about 25 metres away, standing there, panting. He walked within 15 metres of us but was intent on going somewhere else, so luckily we didn't have an incident.

Since then, whenever people discuss how far a lion's roar can be heard I am always reminded of that early morning in the Kalahari with my son. If you hear lions roaring be aware that they're possibly roaring at other lions which just might be right next to you. Whenever we hear lions Adrian always turns to me and says facetiously, "Don't worry Dad, they're still miles away!"

Lions use road networks extensively and are capable of moving great distances in short periods of time. Mostly at night but not exclusively; it's not unusual for a pride to be in the south one day and travel over more than 50 kilometres overnight to be right up in the north the next morning.

Under natural conditions a dominant pride male has about four years before he will be ousted. This is a traumatic time for the pride, and as the pride male ages and can no longer fight off the advances of younger males, the takeover begins. The fighting is brutal and decisive and often leads to the death of the older lion and if not, he escapes to be a nomad for the rest of his life. The new pride male then does something totally contrary to a human perspective – he kills all of the young lions in the pride. When a female is suckling most are not fertile, and with this instinctive knowledge the male kills any cub that is still young enough to suckle. The new pride male's first priority is to make sure his genes are carried on to the next generation, and this is the quickest way to bring on oestrus in the females.

I must say that it is one of the times that I really don't like doing any lion monitoring as it is a terrible sight to see and even with all of the understanding of the life cycle and biology of lions, it is an aspect of lion behaviour that I have not made peace with at all.

It is not long after a takeover that the females come into oestrus, especially if they have lost their cubs, but nature has an interesting twist to the tale because, although mating takes place, the females will not fall pregnant immediately. It gives the females time to see if the new male can defend them from another takeover and that he is tough enough to protect them. It would not serve any purpose to become pregnant if the new male is evicted shortly after because there would be a repeat killing.

The purpose of a lion's mane has been discussed and researched for many years and triggered some heated debates among scientists. It is known that lions in different regions have different mane characteristics, and even back in the early 20th century this trait was linked to a difference in ambient temperature. A classic example is the black-maned lion of the Kalahari. Lions are extremely sensitive to heat and it is known that males in colder climates tend to have bigger manes than those in hot climates.

The mane has often been viewed as a shield that protects a male's neck during fights against other males, but lions mostly attack each other on the back and hips. Instead, the size and coloration of the mane serves as a signal to other lions both male and female about the animal's fitness. By conducting experiments with life-size 'dummy' lions, the Serengeti lion project determined that females prefer males with darker manes. Darker manes also take a physical toll by raising body temperature, meaning that only the 'fittest' males can grow the most attractive manes. The length and darkness of the manes also signal information to other males about a lion's fighting ability. So if we are to accept that female lions are attracted to a darker mane then it would be to the advantage of all males to have a big black mane. However, not all male lions have big black manes yet they survive and breed and become dominant males.

A high volume to surface area ratio means that larger animals have more difficulty with high temperatures. If one looks closely at lion behaviour you will note that a large percentage of their time is taken up trying to reduce heat stress. Some examples are that they sleep in the heat of the day and mostly move by night, lying on their backs to expose their thin-skinned bellies, resting on raised areas in an attempt to catch the breeze and panting after consumption of large meals. Lions do not sweat. Their only means of thermoregulation comes from breathing and radiation from the skin. The mane traps the heat in and so prevents the efficient dissipation of heat. Dark hairs absorb and trap more heat than lighter hairs. The darker surface also absorbs more solar heat than the lighter ones.

Through the use of infrared thermography, which measures the precise surface temperature of a distant object, the researchers of the University of Minnesota managed to get some very interesting results. They spent three months taking heat scans of lions in east Africa. They included the maneless lions of Tsavo in the project. They wanted to see if the difference in body temperature was caused by the difference in body size alone (males bigger than females) or to see if the temperature differences could be accounted for by the presence of the mane, so they needed thermo images of both male and female lions in areas with different temperatures. The maneless males in Tsavo had a similar body temperature to the lionesses in the area, even though they were much bigger. They found that the mane length had little effect; however, males with darker manes were significantly hotter than those with lighter manes, even after controlling for ambient temperature, wind, humidity and prior activity. The same team of scientists also proved that males with darker manes had higher proportions of abnormal sperm. Dark-maned males were also observed to reduce their food intake on hot days. This was not the case with the lighter-maned lions in the same area. So, in conclusion, there may be advantages to having a dark mane but this also comes at a cost, which is in the form of heat stress.

CHAPTER 13

BUILDING THE NEW DAM WALL

After the dam wall was destroyed in the flood, we had some rather interesting times. What had once been a massive body of water had become an area of plenty in the form of vegetation. Of course this had the effect of attracting many herbivores to the area and with the herbivores came the carnivores. The area around the old dam was literally a game drive on its own; everything was there for the guests to see. I was particularly worried about the hippo as their home of water was no longer there. They had moved to a pool below where the dam wall had been and fortunately this pool retained its level even in the dry season. The elephants literally moved in and stayed in the area permanently.

The team of engineers and designers in conjunction with the Department of Water Affairs put together a plan to replace the wall. For my sins I was made the secretary of the Bellair Irrigation Board and was part of the monitoring team when the building of the new dam wall started. The earth-moving equipment moved in and started clearing away the remnants of the old wall. About a kilometre downstream we found a block of concrete which is still there, a stark reminder of what severe damage water can do. The block is ten metres long, four metres wide and three deep. One of the engineers calculated its weight; none of us could believe that the torrent of water had the power to move it one metre, let alone a kilometre.

During the reconstruction some interesting incidents occurred. Because a lot of the game was concentrated around the area, it was inevitable that some kind of confrontation was going to happen. So I gave a talk to all of the dam building staff on what to do in the event of them being confronted by various species. Literally a few weeks later I got a desperate call on the radio that I was needed urgently down at the dam site. When I got there it was total pandemonium – people stuck up cliff faces that under normal conditions would need ropes and climbing gear to get up. Shouting and screaming, lots of waving and pointing at a particular spot under a rock overhang. With all of the confusion I couldn't understand what the problem was and climbed up to see what everyone was going on about. Right there, under the overhang seeking shade, were a lion and two lionesses, relaxed to the point where one female was flat on her back cooling in the shade.

It took a while to sort it all out and get some of the involuntary climbers back to ground level. A few weeks after this incident the ellies decided to take on the role of inspectors of works which created panic and a few incredible athletic feats among the work force. It was really quite funny to walk onto the building site and have a family of hippo literally twenty metres away. Fortunately, there was never an incident with the hippos and as soon as the dam was completed and the water started backing up they moved back to their original home.

The harsh reality of the tension that exists between lions and cheetahs was brought home with a crunch. I was out tracking and kept getting a constant signal from a cheetah out on the dry mud flats of the old dam. I went to have a look for him and as I approached the area there were jackals and white-necked ravens around, which normally indicates some form of carrion. Sadly, there was a large male cheetah, killed by lions. I walked across the area and checked the spoor – what had happened was that the cheetah had taken a springbok and was busy eating when he was taken by the lions. They had ripped him open but hadn't eaten him; but of course, they did take the springbok.

A lady from the village of Montagu sent a baby spotted eagle owl to the reserve. Their scientific name is *Bubo Africanus* and I named this little owl Bubu. Unfortunately Bubu had been imprinted and was totally habituated to humans so it wasn't going to be an easy task getting her back into the wild. Bubu lived with me until she was capable of flying and then I put her in a very large flying area that we had built at the student quarters. She became stronger and flying fit and we released her outside the cage. She never left the area and although she was fed she did become independent. Over the years Bubu became a star in many documentaries and although she was habituated I eventually got her hunting to the point where she could fend for herself entirely.

She was an incredible help to me whenever I gave talks to schools and bird clubs as I used her to show people why they should not have wild birds of prey as pets. What I would do was take her in a covered travel box and leave the door open on the floor facing away from the audience. At some stage during the talk she would hop out and fly across the room, which would get everybody wide awake, and more often than not she would land on somebody's head. I would ask them to just sit still, and Bubu would start preening and I would then explain exactly why wild animals and birds do not make good pets. To this day I get stopped by youngsters who attended my talks and all of them remember Bubu and how she influenced their lives. She was a great teacher who has now gone into retirement and taken up residence in a massive blue gum tree at Gondwana. I went to visit Gondwana recently and had not seen her for four years, so you can imagine everybody's amazement when, as we were walking towards the new buildings, Bubu flew out of a tree and landed on my shoulder. I must say that it was one of those situations where I really found it difficult to hide my emotion. This beautiful owl remembered me and knew exactly who I was after no contact whatsoever in four years!

CHAPTER 14

Some Travelling

Johan and I flew to the UK to investigate the possibility of bringing back a herd of elephants from Howlett's Animal Park in Kent. Adrian Gardiner and Lord Aspinall were good friends and this friendship has continued with Lord Aspinall's son, Damien. So just six weeks after the flood, we flew overseas. We had also arranged to meet with the Born Free Foundation, which has a wonderful facility on Shamwari Game Reserve outside Port Elizabeth. They rescue predators from some of the most extreme conditions you could imagine and bring them to the Foundation to recuperate.

We arrived in London after one of the most uncomfortable flights I've ever experienced. We were given the back two seats on the aircraft and felt every movement it made throughout the night. We were collected from Heathrow where me met up with Paul Gardiner and immediately went to meet the team at the Born Free Foundation headquarters. This incredible foundation was founded by

Virginia McKenna and Bill Travers, whom many of you will remember as the leading actors in the film Born Free, the story of George and Joy Adamson and their adventures with Elsa, a lioness that they returned into the wild.

Their son, Will, had visited our reserve on one of his many missions to Africa and he and I had driven from Shamwari to Sanbona together. During the drive I got to know about the work that his family had started and what he and his team still do to this day. He spent a short time with us on the reserve, meeting our team and discussing the work that they were doing around the world with elephants, lions and tigers.

It was very interesting to see their offices and meet the people who make the Born Free Foundation function so well. They were all very preoccupied at the time with the preparations for a gala dinner to be held at the Natural History Museum in London. It is one of their fundraising events when they raise funds for their ongoing projects around the world. We were invited to this dinner later in the week and I was to experience a most incredible evening.

The discussions in the office were around the Born Free project at Shamwari where big cats that have been rescued by the Born Free Foundation from all over the world, spend the rest of their lives in large bomas under the African sun. A really sad example of the cruelty experienced was an adult male lion that was rescued from a bar. He was found in a cage only slightly larger than himself and was used as a backdrop and entertainment of the drinking public. He spent six years in this confined area before being rescued by Born Free.

After the meeting we went back to London for the night. I was really looking forward to the next day as we would be going down to Kent to visit the trustees and staff of Howlett's Animal Park and Port Lympne. We had been visited by this dedicated group of people earlier on in the year. Lord John Aspinall had created a private zoo in 1957 which had only opened to the public in 1975. Lord Aspinall was known for his hands-on approach to animals in captivity, and over the years had great success breeding with many species such as the western lowland gorilla. They still, under the directorship of his son, Damien, do incredible work with many endangered species. Howlett's had a very good record of returning species back to their home country, which was the whole idea – to get a herd of between six to eight elephants from Howlett's back to South Africa and onto Sanbona.

We were there to discuss the logistics with their team. On arrival, we went for tea and refreshments with Damien. After we sat down a butler walked in, or should I say 'glided in.' He avoided all eye contact and without saying a word he ascertained what beverages we wanted and glided out again. I was absolutely fascinated – here I was, an African, sitting in a stately home in England with a butler serving tea! I nearly started giggling. I looked across at Johan and Paul and they were just as mesmerised as I was; in fact Johan had a very perplexed look on his face. I had only read about people having butlers. When he glided back in, it was with a silver tea service that I'm sure must have been worth a small country's GDP.

Damien is a very down-to-earth 'hands on' type of person and is absolutely passionate about what he does. He briefed us on what was going to happen over the next two days and off we went to see the layout. We were going to stay in the area for two days and visit both of the Parks and get the opportunity of seeing their gorilla programme and talk to the team that did the reintroduction programme.

We spent the afternoon with the elephants and discussed all of the logistics of crating and transporting them, finding a big enough aircraft to fly them and all of the details that would go with being a crew member on a flight containing a herd of elephants. Even down to getting feed from South Africa six weeks before the flight so that they would already be accustomed to what they were going to eat when they got here. That evening we sampled some of Britain's beers and I must say that for the first time in a couple of days I slept like a baby. We were there in May and during the course of the next day we had light snow, heavy rain, sleet, fog, drizzle and in the afternoon bright blue skies with sunshine – and this is what the English folk call Spring!

The next day we went down to the sister animal park, Port Lympne. It was here that I really was encouraged to see what these good people have done for African wildlife. Over and above the elephants, they offered us six black rhinos to take back with us. I couldn't believe their generosity. The rhinos were having their feet checked when we arrived. The team is positively inspiring, doing some remarkable work behind the scenes, training the rhinos to enter a specific area so that they can work on them without using tranquillisers or any form of sedative. Seeing a small man working on a massive black rhino's foot is the last thing I was expecting to see. He was chatting away to the rhino and working on its feet as though they do this on a daily basis.

We then had a very special treat – a behind-the-scenes look at the gorilla project. Having, at that stage, not seen a gorilla in the wild but only in photographs and past zoo visits, I had a good idea of the size and beauty of these great apes but nothing really prepares you for a close-up encounter. As we rounded a corner in the building, there, right next to us was a massive silverback.

We had been briefed before the time and advised not to look straight into the gorilla's eyes, but this chap looked at us in a way that humans look at each other. He was part of a breeding family from which young members are selected and sent back into the wild. We got the opportunity to climb to the top of the cages to observe the gorillas from a height. They have a very well-structured family hierarchy with the silverback having the final say in most of the domestic disputes. As big as they are, they're capable of the most gentle acts; often some of the youngsters would huddle up to their father and he would hold them tenderly and soothe them with gentle noises. One can see the respect and love in their eyes and actions. Further out in the park is an enclosure where the public can observe the gorillas through strengthened glass while they go about their daily lives.

The Aspinall Foundation manages and supports a number of conservation projects worldwide to help preserve and restore wild animal populations. They have projects in Madagascar, Indonesia, Cambodia and gorilla programmes in Gabon and Congo. It is feared that the western lowland gorilla will be extinct by 2020 if their numbers continue to decline at the current rate, mainly due to deforestation, the Ebola virus and the bush meat trade. The Foundation has had enormous success reintroducing gorillas back into the wild in Gabon and Congo.

We were taken on a quick sight-seeing tour and went to have a look at the White Cliffs of Dover, a truly magnificent geological feature that is full of military bunkers – a stark reminder of what the British people went through protecting their shores. Everywhere we went there was the most amazing history, castles and churches dating back tens of centuries, even some of the roads having been built by the Romans. The road between Howlett's and Port Lympne had once been the haunt of a famous highwayman called Black Robin. (I have since found out that a lot of highwaymen were called Black Robin but it didn't change the mystique of the area.)

On our return to London we had the opportunity of experiencing why the people of London call their ring road a parking lot. Justin Timberlake was going to give a concert, so all driving came to a stop. It was quite amazing that one man and his band could bring the whole highway network of a city the size of London to a complete standstill. It took us ages to move one kilometre. What was supposed to be a two-hour journey took just over seven hours. I'm surprised that people don't expire in their cars – you can imagine what would happen on a hot day if you didn't have aircon.

That evening we attended a function held by the Born Free Foundation at the Natural History Museum. Johan, Paul and I had to wear dinner jackets and felt strange as this is not your regular bush wear; we really felt out of place to start with. We were met by two rows of absolutely gorgeous Indian dancers from a play that was on at the time. Inside, Virginia McKenna welcomed us (she and my mother could be related, they look so similar), a very gracious lady indeed and it was a privilege to meet her. Before we were seated a very glamorous lady approached us, rather tall and with all of the right parts in their respective places. Johan couldn't resist and asked her if he could have a photograph with her – it turned out to be Rachel Hunter. I had no idea who she was but I have since been informed – by my wife – that she is a famous model who was once married to Rod Stewart.

As I was about to sit down I saw a lady and gentleman gesticulating at me. I recognised the man as Martin Clunes so I went over to them. Somebody must have briefed him about the dam failure at the reserve because he knew all about it and was very sympathetic towards us. He is rather tall and has a wonderful sense of humour, and was the master of ceremonies for the evening, keeping his audience in fits of laughter. The evening was a great success for the Foundation and I enjoyed every minute of it. I got to meet many wonderful people that night including Brian Adams and I am still in contact with some of them to this day. I don't think my family believes that I met some celebrities who actually wanted to speak to me!

CHAPTER 15

Aardvark Holes

The flight back to South Africa was a wonderful experience, the opposite pole to what we'd had going there. Some very kind fairy godmother was looking after us. When we checked in at Heathrow, the lady behind the counter checked our paperwork then looked up and, with a smile, told us that we must have some good friends in high places because we'd been upgraded to business class. And not only that but that we could use their VIP lounge before the flight! My word, what a difference it was from being at the back next to the toilets. As the aircraft took off the gentleman in the seat next to me asked me to wake him when we arrived in Cape Town. I thought he was joking but he swallowed some pills and that was the last thing he did for twelve hours till I shook him awake as we flew over False Bay for the landing. Johan flew on to Port Elizabeth and I drove back up to the reserve. I was to visit England again later that year, this time both for work and pleasure.

Once back on the reserve the team brought me up to date with what was going on and I gave them a full report about the trip and the possibilities of getting the elephants and the rhinos. The team had done a great job in my absence and all was well on the reserve. The lions were fine and the cubs were growing up nicely. We were under pressure to allow guests to visit the cubs so we had a meeting with the rangers and agreed on the best method of viewing the cubs without habituating them to humans. The rangers are a dedicated group of people who love the animals and plant life and are capable of communicating their passion to visiting guests. They soon had the hang of it and many guests saw the project right from the start and were privileged to get the explanation from the enthusiastic rangers.

Early one morning I was observing the cubs and it became obvious that all three were males. When they're born it's difficult to sex cubs; but as these got to about six weeks the various apparatus become obvious. After I'd had time to reflect on this, I realised that it wasn't a bad thing at all. If you want to breed white lions then you need white lions. If you want to get the gene back into the wild then what could be better than three males with the gene. The idea from the start was to get the gene out there, so that all of the white lions that had been removed from the wild could be replaced with lions that carried the gene and in the course of time the possibility of a white lion being born in the wild could happen. With three healthy young males we could introduce them individually to three separate prides of females or even keep them as a coalition to a single pride. We now had options that we could work with. The area around the boma was active with many game species, so the cubs got to see gemsbok, zebra, eland, springbok and other prey species to stimulate their curiosity and make them aware of their prey items.

They were being fed from behind a system of covers so that when Queen brought them the carcass they couldn't associate it with humans. I was very concerned about this aspect, as I was sure that even though we were doing our best to hide from them the fact that Queen and Jabulani were being fed, I wasn't convinced that the cubs hadn't picked up on what we were doing. I was very encouraged when after one feeding I witnessed one of the cubs with its mouth around the throat of a springbok that had been fed to the lions. This is the killing method that they use and if the cubs were practising on an already dead animal, that was a big step in the right direction. In a normal pride situation, the young males would be ejected from the pride by the dominant male when they reached the age of about 20 to 24 months, so we still had time to work with the cubs before we needed to remove their parents.

On the main reserve we had introduced three more sub-adult lions right down in the southern sector at Old Nick's dam, which we named after a massive territorial mountain tortoise with a nick out of his shell. The idea was that the reserve was more than big enough for two prides so we would have one in the north and one in the south. The pride in the north was now a stable and functioning unit with one adult male and two adult females; a lot of mating had occurred so we knew that pretty soon we would see the arrival of the first cubs.

Nothing it seems happens without some or other 'incident' taking place. We got a call from the rangers that there were only two of the newly introduced lions in the boma. Before they were put in we had put tracking collars on two of them, so down south we went with the telemetry kit to see what had happened.

When we got there, sure enough there were only two lions in the boma, one with a collar and one without; so luck was on our side, we could track the missing animal. I tried the telemetry and we received a strong signal from one of the transmitters and nothing from the other. We then walked around the boma and started to get a faint signal from the other collar. It appeared to be coming from the centre of the boma yet we could only see the two we'd seen already. After going right around the outside perimeter and being certain that the signal was definitely coming from its centre, we were completely perplexed. The boma was built in a very low scrub area and we'd even had to build a shade structure for the lions to lie under during the heat of the day. The dam wall was next to the boma so we climbed up and took turns looking through binoculars to see if we could see anything. We then sent teams up the ridges in the area to see if we could triangulate the signal.

All of our efforts confirmed that the signal was coming from right in the middle of the boma. I climbed up the dam wall with a pair of binocs and noticed that there was an aardvark hole right in the centre of the enclosure. This was the only place the lion could be! We looked at each other and decided we needed to drive inside to have a closer look. We put the students on a vehicle on the western side of the boma with instructions to keep us up-to-date about what the other two lions were doing, and to give us a sort of running commentary on their movements. We had the dart gun and the medical box with us so if we needed to medicate we wouldn't have to leave the boma. Sure enough, there in the hole we could see the side of a female lion. She had gone down the hole and when she tried to reverse out she'd put her backside into a branch tunnel, effectively pinning herself in the hole.

Then, to add to her troubles, the walls had collapsed around her and disabling movement. We could see her mid-section and one paw and thankfully she was breathing so we knew she was alive, but the problem was that at that range it wasn't possible to dart a lion with the type of gun we had. A pole syringe would have been ideal if only we'd had one. The other two lions were a bit antsy but while we were in the vehicle they weren't too aggressive or under stress. We had to do something quickly, so I got the vehicle as close as possible to the hole without the possibility of making it cave in any more. Paul filled up a syringe with tranquilliser and very slowly opened the door to get out, but then the two lions let us know that they didn't approve of our presence and gave us quite a few hairy moments. The female was stuck at least a metre down the hole so Paul had to literally crawl down to get the injection to her. The main problem was the paw – if she heard Paul approaching she could easily tear him to shreds.

The other major concern was that if she somehow broke loose while he was down there, I didn't even want to contemplate what could happened. All the time this was going on the other two were growling and making aggressive charges with snarls that demanded extreme levels of nerve and concentration. Paul wrapped his jacket around his left arm, putting the syringe in his right hand and slid into the hole. The female sensed him and lunged at him with her paw and hooked into the jacket. He then successfully injected her just above the paw and had to wait until she relaxed her grip on him to wriggle out of the hole. After he was safely back in the vehicle we let the tranquilliser do its job but waited an extra few minutes to be on the safe side. We then had to dig her out of the hole without her suffocating and the other two lions causing any of us injury. We shovelled a whole pile of soil while the students kept up a running commentary about what the other two were doing. Even though they were sub-adults, the growl of an angry lion gets your absolute and undivided attention and does nothing for your digging skills. Because we didn't want the soil to fall back down the hole we had to remove a massive amount of it, taking ages because it was a compacted clay and incredibly resilient. Eventually, we freed her and were really happy that she was none the worse for wear.

Before we left the boma we filled in and compacted that aardvark hole so as not to have a repeat of this exercise. Every boma that I have been involved with since has been thoroughly inspected for any aardvark holes or any possible tunnel that an animal could crawl down. I get the cold shivers when I think back on this incident, Paul sticking his whole body down a hole to inject a trapped lion. This is true bravery, and the mark of a man who really loves the animals he works with.

Jabulani and Queen.

Queen and the cubs at two days old.

Queen presenting the cubs in the open for the first time.

Confirmation that they are all males.

Queen on her look-out dune.

Together as a family for the first time.

Doing some exploring away from the brothers.

Displaying the killing bite at a very young age.

The release day.

Queen returns to collect the reluctant cubs.

TOYOTA
TOYOTA
TOYOTA

64 and cubs, the first wild born Cheetah in the Karoo for hundreds of years.

Two of the youngsters showing their manes and the father of the cubs stretching in the sunset.

White and Tawny females.

Author and a sedated juvenile.

Jabulani

CHAPTER 16

More Lion Problems

The reserve was becoming busier and busier with more and more guests arriving. It was very rewarding to see the reactions of the guests on their game drives. Often I would be out in the late afternoon working in one spot or another on the reserve and if I was doing any monitoring and had any of the cats or elephants nearby I would tell the rangers about it over the radio and they would join me, or I would get away before they arrived.

One afternoon I was down in the south monitoring the cheetahs and checking on the cubs when a game drive drove past without seeing me as I had parked my vehicle behind a dense clump of reeds. Once I'd found the cheetahs and checked that all was well, I radioed the ranger to bring his guests to the animals. I stayed with the cheetahs and when the guests arrived the ranger gave them a small briefing on what was expected of them when they were in the presence of cheetahs. He brought them on foot to where I was waiting for them.

I really enjoyed their reaction, because as big as cheetahs are they blend in so well with their surroundings that if they lie still you can't see them. So when I pointed them out all I got was blank faces; I then realised that they were looking into the distance. When I told them to look just six metres in front of them the reaction was an absolute treat! They couldn't believe their eyes; lying literally in front of them was a female with four three-month-old cubs with tummies full of springbok that the female had hunted earlier. I cherish those moments and still to this day love just sitting quietly out in the wild with cheetahs; they are so regal and yet amazingly tolerant of humans who treat them with respect.

I have a passion for birds of prey, better known as raptors. Many years ago while I was working on vegetation evaluation plots I started to hear a specific bird call. It was the pale chanting goshawks vocalising to each other. I started to listen carefully and eventually could tell that they had different calls for different situations; for example, if another bird of prey was in the area they would give off a very fast high-pitched call. On our farm a pair nested very close to the house so I had the wonderful opportunity of following their lives closely. They were usually the first birds to call in the morning and they would often roost in a tree near the house and be my morning wake-up call. The pale chanting goshawk now known as the southern pale chanting goshawk – and from now on in this book as a PCG – is a medium-sized raptor about 55 cm tall with a wingspan of 110 cm. As per the norm in raptors the females are heavier than the males, coming in at about 1 kg whereas the adult male is about 750 g. *Melierax canorus* is its scientific name. The derivation of the scientific name refers to its melodious song; *Melos* is Greek for song, *Hierax* is Greek for hawk and *Canorus* is Latin for melodious, so it is aptly named. Although the books say that it is more often than not a quiet bird, I haven't found this to be the case. I have seen and heard them call in many different circumstances, from a stable hunting platform to a wobbly branch, while flying and also while on the ground.

Their display flight is a beautiful slow beat of the wings while rising up and stooping down in perfect formation, all the while calling to each other. During the course of my studies I came to the conclusion that the PCG territories and nesting intervals are closer in vegetation that is in good condition, rather than that of overgrazed areas. I won't go into too much science here but what this means is that you can use the PCG as an indicator species in the Klein Karoo because if the vegetation has in the past been over-utilised, the territories will be as much as 64 square kilometres whereas in vegetation that has been used prudently, the territories can be as small as four square kilometres. Of course there are other factors that come into play but I often drive through an area doing a raptor count, and if the PCGs are far apart you can bet your bottom dollar that the area is in bad shape.

Nesting is done in the colder winter months but I have seen breeding taking place from June right through to April with the peaks in the Klein Karoo around August, September and October. Mostly two eggs are laid; I have seen three but that is exceptional. The makeup of the nest is really interesting, mostly a base of twigs then a veritable Pandora's box of paraphernalia on top. Zebra dung, bits of grass, sheep fur, old rags stolen from washing lines… I even witnessed a PCG returning from a foray with pink roofing insulation that it had collected from a building site five kilometres away. To do this it had to fly through three other PCG territories, and they are not very tolerant of each other and will defend their territories aggressively.

At many of the nests that I monitored I would find that the nest had been lined out with spider web. After some careful searching, I found it to be from the social spiders' webs or retreats as they are now known. Stegodyphus spiders live in colonies and their webs are very large. The PCGs collect the web and line their nests with the silk; I also suspect that this helps the social spiders disperse to create new retreats. In nature so many inter-dependencies exist; if one aspect of an environment is removed the whole environment is compromised.

There are two places on this planet that I really love, the first being the Klein Karoo and the second the coastal village of Betty's Bay. Simmy and I have built a bungalow at Betty's Bay to retreat to when we need to regain our senses. It is a truly beautiful part of the world where there are no street lights and exterior lights are not allowed to be left on all night, allowing for some of the best star-gazing evenings I've ever had. On one of my few off times Simmy and I went down for a few days getaway, and we had just arrived at the bungalow when a call came in from Ryno. The bad news was that the big male lion in the north had escaped and was in our northern property that didn't have a game-proof fence on its northern boundary.

I wanted to leave him to get it sorted out as he was more than capable of handling the situation but the ramifications if something went wrong were just too much to contemplate.

The neighbour to the north is a good man but the idea of a fully grown male lion wandering around his livestock would test any farmer's sense of humour to the limit. So I asked Ryno to take the team and the dart gun and see if he could get him, and in the meantime I would be on my way. Fortunately for me my wife is a very understanding person and we packed up and left for the reserve very quickly. It was about a two and a half hour drive and they kept me up to date via my mobile phone until I was within radio distance. The lion was trying to get back into the reserve but because we had habituated him to the electric fence he was reluctant to approach it.

There was no choice but to dart him, but unfortunately he had climbed a ridge which our vehicles couldn't get up. We had to circle a long way around to get up on the ridge as it was just too dangerous to approach him on foot as he was already stressed beyond normal limits. Fortunately, he was wearing a tracking collar so we could keep tracking him even when we lost sight of him. He would walk in one direction and then just as we were set up and ready, he would change direction, so it was a very frustrating set of events. Ryno and I went in every direction to try and dart him but he gave us the runaround at every opportunity.

It took most of the day but eventually just before dark we darted him, and when he went down we had the opportunity of giving him a good inspection and could also look for any clues as to why he'd escaped. No injuries whatsoever could be found on him and he was in excellent condition, and because there were no other males on the reserve that he had come into contact with he had no scars to characterise his face, so he even looked handsome. We got him back into one of the bomas to wake up. It's always best to let a lion come around from the tranquilliser in a safe environment because if there are other lions in the area they'll attack the semi-conscious animal because to them it's behaving in a strange way. Also, they are vulnerable to being attacked by other predators like hyaena.

We checked the fence line the next morning to find where he had escaped, but it was an absolute mystery. We could find no track or trace of where he went over or through the fence; it remains one of those mysteries to this day. After making sure the fences were 100 per cent, we let him out and he never gave us a problem again. I had to report all of this to the bosses and the nature conservation officials and of course we played it down to the best of our ability, but even though we'd got him back fairly quickly it wasn't a good thing to happen. We were trying to get permits for various other animals and the fact that an escape was possible wasn't a good sign. The saving grace was our quick actions once we realised that the animal had got out of the fenced area. We had all prepared for the moment and an action plan was in place, but wild animals are unpredictable and once they escape and are under pressure they become erratic, all normal behaviour flying out of the window. And when you're dealing with a lion it becomes rather dangerous to say the least. The average Joe public doesn't encounter lions in his everyday activities and when they do, it can be quite an event which could end up with either a very relieved smile or a trip to the hospital.

Our team was made up of farm workers who'd never even seen some of the animals we were introducing, let alone handled them. But it must be said that of all of the animals we introduced the lions had the most galvanising effect on them. We had given them a lot of training about what to do and what not to do and they handled themselves with distinction. After the dreadful tragedy of the dam breaking, all of the staff who had lost their homes below the dam were moved up to the new wildlife office area.

I was allocated a small house to use and before the fences were erected the ellies would often visit at night. I love chillies and it appeared my predilection was shared by elephants as every time my chillies were ready to be picked they would simply disappear! One night, after hearing a loud crash and shouts, I ran out to nearly get flattened by a fleeing ellie that had just received a tongue-lashing from Barnard, who is the most retiring and quiet person you've ever met. That is, until his vegetable garden comes under attack.

So the staff were becoming used to the idea that all of the domestic farm animals were a thing of the past and that wild animals had taken their place. I later found out that Barnard had watched me stand down the elephants one night when they had entered our compound. According to him, I'd spoken very nicely to them until they didn't listen, then I shouted at them and slapped my hat against my thigh, which he told me got them to move out of the gate which some twit had left open. And this was the method that he'd used to great effect that night as his veggie patch disappeared down the throats of the ellies.

At all of the introductions we would involve as many of the staff as possible; this also included the hospitality staff if the timing was right. All would get a briefing on the specific animal to be released or put into a boma, and what benefits it would have to the reserve and why the area had suffered without the specific animal's presence. They all knew about how many of the tree species had stopped flourishing because there were no elephants and it was a very rewarding day when one of the chaps pointed to a piece of ellie dung – and there was a small tree growing out of it.

CHAPTER 17

LION GENETICS

This is the easiest way that I can explain them, with just a tiny twist in the tail.

Fossil evidence suggests that lions have roamed the earth for about 1.8 million years. They were formerly the most widespread large land mammal on the planet. Lions ranged throughout Europe, Asia, the Americas and Africa. It is thought that the decline in their distribution was due to expanding human populations and climate change that occurred at the end of the Pleistocene era which caused a mass extinction of large mammals.

Currently, wild populations of modern lions reside only in Africa and Asia with both populations becoming increasingly endangered. They are listed as vulnerable and endangered on the International Union for the Conservation of Nature (IUCN) red list. However, it should be noted that up to 24 subspecies classifications for lions have been identified over the years, based solely on geographical and physical attributes like mane size, body size, mass, skull size and colour. Eight subspecies are recognised today; however, due to the subspecies classifications being based on non-scientific evidence, the IUCN has documented only two subspecies of lions, after one of the earlier genetic studies made the distinction between African lions and Asiatic lions.

Angola Lion *Panthera leo bleyenberghi* which occurs in Zimbabwe, Angola and Zaire.

Asiatic Lion *Panthera leo persica* from the Gir Forest Sanctuary in Northwest India.

Barbary Lion *Panthera leo leo* from North Africa – this species is extinct; however there is a possibility that there are some in captivity.

Cape Lion *Panthera leo melanochaitus* from the Cape Province, but now extinct.

East African Lion/Masai Lion *Panthera leo nubica* sometimes referred to as massaicus from Kenya and Tanzania.

Senegalese Lion *Panthera leo senegalensis* from west Africa to the Central African republic.

Transvaal or South African Lion *Panthera leo krugeri* from Botswana, Namibia and South Africa, including the Kruger National Park from which it derives its name.

Recent genetic studies of lions have provided support for some of these earlier classifications and this shows how important genetic research is. Genetic evidence suggests that the Asiatic lions' ancestors split from the ancestors of the Senegalese lions between 200 and 70 thousand years ago. Don't you just love how much variation there is in the timing of this event?! It is also interesting to note that while the Cape lion had similar physical attributes as the

Barbary lion, DNA research doesn't support its status as a distinct subspecies. It would appear that the Cape lion was made up of the southernmost population of the Transvaal or South African lion *Panthera leo krugeri*. Sadly, the Cape lion became extinct in the wild around 1860.

There are clear genetic differences between the Asiatic and African lion, which supports the subspecies designation of *Panthera leo persica* and *Panthera leo leo*. A strange fact though is that lions from west and central Africa show a remarkably closer genetic relationship to lions from India than to those living in South and East Africa. Not only is the physical appearance of the Asiatic lion different from the African lion but the genetic makeup also differs. Asiatic lions are thought to have low genetic variability in contrast to African lions. African lions' DNA is said to be heterozygous which means that they possess two different forms of a particular gene, one inherited from each parent. The Asiatic lions are homozygous, which means that they have two identical forms of a particular gene, one inherited from each parent.

Studies have provided evidence that the genetic diversity within lions might be more than just the two subspecies that the IUCN recognises. So why all the fuss, you might ask? Well, it's quite simple really. Let us say that, for example, the lions from Angola are nearly extinct, but because they don't have the status of a separate lion sub-species, they are then not recognised as endangered and fall into the same category as all of the African lions. This makes conservation efforts in certain parts of Africa very difficult indeed.

Lion populations are becoming isolated and many are confined to protected areas that have a limited gene flow.

Managing these populations appropriately is vital in order to minimize loss of genetic variation. Translocating lions can prevent this loss and ideally should mimic natural gene flow by transferring those from a nearby area that are genetically similar. When lions are reintroduced to an area, or a population is restored in depopulated areas, only animals with a similar genetic background should be used. Lion conservation will benefit from a better understanding of lion subspecies, and it will be a great step in the right direction when conservationists begin to recognise the different genetics of lion populations.

Having said all that, it is not at all wise or even possible to just create your own version of a subspecies. I have just read in a magazine about a new species of lion called *Panthera Leo Krugeri rara avis.* In Latin *rara* means rare and *avis* means bird. On further reading it became clear that this is what they were calling the 'white lion.' Some person or persons without the necessary scientific confirmation and input are attempting to create a new lion subspecies.

I have no doubt that the white lion is a rare animal indeed. On further investigation, the use of the term *rara avis* means a rare or unique person or thing. Now we all know that white lions have appeared throughout Africa on very rare occasions. This is part of their genetic makeup and this is what makes them so very special. However, despite loving these animals with an absolute passion, I have no desire whatsoever to have them become a subspecies. They are neither a person nor a thing; they are white lions and they deserve their place in the African landscape alongside their tawny brothers and sisters. At the beginning of the book I gave a description of what a white lion is and I repeat it here so as not to create any confusion. A white lion is a lion that has a recessive gene – the condition is called leucism.

CHAPTER 18

THE RELEASE

The white lion cubs were doing well and displaying normal behaviour, chasing their parents' tails and one another, in circles. Stalking any insect or lizard that happened to crawl past … even small rodents were pounced upon. The idea was to open the gates when the cubs were big enough to keep up with their parents, and give them the opportunity to get to about two years of age in the most natural environment possible. The release date was set and the preparation and planning for the event began. It was going to be a very special occasion; the first release of white lions into the wild was finally a reality. All of the hard work to date was going to be rewarded. The possibility of getting the gene into the wild was now a reality; the work, however, was far from over and years of frustration and anguish lay ahead. But a very big step on the road was about to take place.

The day was fast approaching and the team was briefed over and over again in preparation. I had to answer many questions from the bosses, and the press also wanted information. It was very important for the project that things should run smoothly as the eyes of the world were upon us. One small slip could really upset the project's future; however, all the preparation in the world cannot predict what a wild animal will do at a certain time and under specific circumstances. In fact, predictably unpredictable is about as correct as you're going to get. For some reason or other I can't actually remember the reason why, it was delayed for a few days. I know there was a mad scramble to contact the necessary people so that they could change their schedules.

When the new date finally dawned we met all of the guests and press. Gaston and his family were there, it being a very special day for him as it was his idea in the first place. We gave a briefing on what we'd achieved to date, and then everyone climbed into a convoy of vehicles for the drive down to Brakrivier where the team had built a small fence, creating an L-shape design from the boma, as well as a platform from where the photographers and camera people could operate. Before the folk arrived Ryno and the team had culled a kudu and pegged it down about 30 metres out of the gate so that the lions would be enticed out of the boma on their first steps to freedom. I took on the responsibility of opening the gate, which sounds like an easy thing to do, but when a gate opens inwards and there are five lions on the other side of it who can see a delicious meal just thirty metres away, it can become problematic.

I had a cameraman in the back of the vehicle with me and the rest of the team were in place behind the fence. It was time! A few days earlier I had rigged a gate-opening device that would work perfectly, I thought. If I could push the gate back about a metre, I could then drive away to a point on the fence where the rope could be accessed and I could then just pull the gate open all the way. The best laid plans of mice and men... We had checked the system when the kudu had been pegged out that morning and it was all 100 per cent.

When I got there to open the gate, the whole system had been trashed. I suspected that the cubs had found it and destroyed the whole device. I had no choice so with cameras on me, and Jabulani, Queen and the cubs looking at their next meal directly behind my shoulder, I slid out of the vehicle and pushed the gate back as far as possible without alarming the lions. I then got back in and drove away to a point where the cameraman in the back of the vehicle could get the footage he needed.

Jabulani was the first one to start walking out; it was a breathtaking sight to see this magnificent adult white lion majestically stride straight out of the gate and walk over to the kudu and start eating. Queen followed shortly afterwards and joined him. The cubs were shy and stayed behind and after a while Queen went back to the gate and called for them. Two of them followed her but the last one just lay down inside the gate. Jabulani, Queen and two cubs eating away on a carcass was a very special sight to see, but the last cub didn't move. He stayed lying down behind the gate, when out of nowhere a dust devil appeared and they all got blasted by the sand. Queen then got up and walked back into the boma and the last youngster followed her. It was a great relief to see the last youngster finally join his family.

The photographers got a second chance to see her walk out of the gate with a cub in tow. The opening device that we had created was also a closing device as I wanted to close the gate once they had left the boma to prevent them from going back in. However, in retrospect it wasn't a bad thing that we couldn't close the gate immediately because they still had to venture away from the boma to find the water sources that had been prepared for them. So in the meantime till they found them, we would leave the gate open so they could access water from a source that they knew if they needed it. The day was a great success for the team and I really enjoyed the fact that they received the accolades they did. The whole team had worked incredibly hard to get the project to where it was and to this day I am very proud of each of their contributions in making this happen. They can truly say that they were part of the team that returned the first wild white lion back into the wild. Gaston's contribution to this project can never be forgotten because it was his idea to start with, and people can say what they like but the first white lions that would make a positive contribution to restoring the white gene back to its rightful place was done on Sanbona in the Klein Karoo. Having said that, the work really only started on that day and there were many years ahead before the project would really show us that we had made the right decision.

CHAPTER 19

A Sex Shop and Mountains Walks

About 25 kilometres out of Barrydale on the road to Ladismith is one of the world's most talked-about roadside landmarks. On the right hand side of the road set in the dry arid scrub of the Klein Karoo, surrounded by a couple of exotic trees, is an old homestead with some red writing on it. As you approach, the writing becomes visible to reveal 'Ronnie's Sex Shop.' I can assure you that this is possibly the last place on this planet that you would ever expect to find a sex shop, and if your curiosity gets the better of you, you may or may not be disappointed to find that it's not a sex shop after all, but a small pub. Many years ago, Ronnie, the owner, moved away from the hustle and bustle of the city wanting to start a shop of sorts. He had painted Ronnie's shop on the side of the building and one weekend a couple of his friends added the word 'Sex' in between the two words, and the rest as they say, is history.

It's one of those places that you really just can't drive past, and has become a tourist must-see. The pub is small and welcoming with some rather interesting décor. The area behind the bar is made up of a wall of knickers of every description and other paraphernalia, and the ceiling has a collection of peak caps that have arrived from around the world. Ronnie is a wonderful host and has been a supporter of our reserve since inception. Ronnie's is also strategically situated about a kilometre before the turnoff to the northern sector of the reserve, making a stop for some form of libation on the way home a must.

For the students there were two excursions that became a tradition. I think there might well have been others, but they don't bear contemplating by a man of my age. The one was a braai (barbecue) at Ronnie's and the other was a walk through the mountains. Separating the north from the south of the reserve is a mountain range that runs in roughly an east-west direction. Known as the Warmwaterberg (Berg is Afrikaans for mountain) for the hot springs that occur in places, the range is made up of Table Mountain sandstone with many quartz infusions and is on average about 1,500 metres high. These beautifully-folded mountains were created millions of years ago when South America pushed up against the African continent. Over the millennia erosion by wind and water has created kloofs in the range and there is a specific one which we named Leopard's Kloof. (Kloof is the Afrikaans word for a valley or canyon.) Leopards Kloof is approximately seven kilometres long and very narrow.

Being in an arid zone, the little river only flows after heavy rains and for the rest of the time is dry with just a few rock pools that retain water all year round. The walk normally starts on the southern side where the beautiful folds are clearly visible. About halfway up the ridge on the east side is an overhang which I climbed up to see if there was any rock art. Sadly, I found no paintings but did find a ladder made entirely from wood, up against the side wall of the overhang. It was incredibly well crafted with no nails or any metal whatsoever being used. It was really puzzling at first why anybody would have dragged this heavy ladder halfway up a mountain, but on further inspection I found a beehive.

Some person from hundreds of years ago had been harvesting this honey in such a way that it is still a viable beehive to this day. As one follows the canyon it starts to snake left and right with very steep walls, and in some places they are so sheer that you have to swim across the rock pools. This is wonderful in summer but in winter it can take your breath away. There are many great spots along the way for a picnic, and one of my favourites is a place where there is a massive wall of rock, and situated close to the top of this sheer cliff face is a nest which is used by a pair of booted eagles. This in itself is a rare occurrence because the booted eagle is a migratory species and normally doesn't breed when visiting this part of the world. In early summer they arrive, when the male starts beautiful display flights over his territory; calling, he rises high up on the thermals and then literally plunges straight back down at high speed and then rises up steeply again, almost like a rollercoaster. After the nest has been repaired and the female is incubating, you can often see the male hunting and delivering food to the nest in the form of mice and lizards.

The walk is not far in distance, but it covers rough and uneven terrain and normally takes about four to five hours. If you don't have a certain level of fitness it can leave you rather tired, but elated at achieving a milestone that will be embedded in your memory for life. For the students it is a badge of honour to say that they have walked Leopard's Kloof and for the person who walked with them it was great to see youngsters absorbing themselves into Mother Nature. To celebrate the walk and life, a braai at Ronnie's is a must. This involves the student co-ordinators packing up all the necessary goodies; normally meat or chicken of one or another cut with the good old South African stalwart, boerewors, a delicious spicy sausage that is cooked over the coals in a large coil. Then some salads and bread and a couple of spices and everybody gets into the vehicles and travels over to Ronnie's, which is about forty kilometres away.

Many drinks will be enjoyed and a great meal eaten, and then the youngsters will party the night away, dancing and singing till the early hours of the morning. In South Africa a popular libation is brandy and coke. My personal tipple is normally either red wine or a whisky; however, it is absolutely mandatory for me to have a brandy and coke whenever I visit Ronnie's. The coordinators have to be stone cold sober to make sure that the students get back to the reserve with no other affliction than the forthcoming mother of all hangovers awaiting them.

It never took very long for a student to get a nickname, especially if they had some interesting quirk or obvious feature. We had a Sally Suitcase, something to do with chocolate, I think. There was Pocket Size, because of her diminutive stature; this name was then used on anything small, including baby elephants. Suzy became Pocohontas as she had long plaits, Malcolm became the Bushveld Lothario and many others acquired some rather interesting names.

Once a month the local district nurse and her assistant would arrive to do health checks in the district. This normally involved checking general health, pregnant mums and newborns. A very important task indeed and these two wonderful nurses took their work to heart. They had the total respect of the community from a professional level and the fact that they were both rather formidable ladies. When they alighted from their mobile clinic you could almost hear it give a sigh of relief and actually get taller. However, they were gentlewomen and most of the babies that arrived in the district were delivered by these wonderful ladies, whose no-nonsense approach to life was quite refreshing. With them a spade was a spade.

After a discussion with the team it was decided to get a supply of condoms for the student centre; so when the district nurses arrived I invited them in for tea and started to ask as delicately as possible for a box of condoms. As soon as I started talking they picked up on what I wanted without me having to finish the sentence. I asked them to please drop them off with Karlien at the student centre, which they were happy to do as it was on their route. About an hour later a rather flustered Karlien walked into the office and explained to me that while she was showing some VIPs around the student centre the district nurses arrived and shouted across the road, "Hello Karlien, we've brought you your supply of condoms." Karlien had cringed and you can imagine what went through the VIPs minds when the nurse then produced a massive box of condoms from their mobile clinic!

The reserve was never without some form of orphaned or injured animal that had been brought to us by a kind person who had rescued it in the district. Pete and Nola Fraser, who had an animal rehabilitation centre in the district, arrived one day with a pair of juvenile Cape clawless otters. Knowing that our dam supported a healthy population of otters, we accepted them and started their introduction process immediately. In front of the new wildlife offices the builders had reconstructed an old earthen dam which was about 30 metres long and 10 metres wide but at least six metres deep, ideal for a pair of teenage otters to learn how to catch crabs, frogs and fish and whatever else was on their dietary requirement list.

We put them in a very large cage next to the dam for a few days so they would know where to return to get food if required. Then on a Saturday morning when the office and living quarters were quiet, I opened the door of the cage. It was very interesting to watch them play and shuffle around but they didn't get in the water, so I went and put on my swimming trunks, grabbed some of their food and dived in. The reaction was instantaneous; they dived in and swam circles around me all the while playing and biting at the food. Ryno had put some fish in this dam to keep the mosquito population at bay but at this age play was the order of the day so their hunting skills were being learnt using the 'play time' method. Otters are part of the Mustelidae family, along with weasels, pole cats and honey badgers. All are known for their exceptionally strong jawbone structure and formidable canine teeth, which can inflict serious injuries on anything foolish enough to harass one. Once they set their teeth into something it is almost impossible to dislodge them. Many a visitor to the wildlife office got ambushed by this pair of ravagers. I can tell you that it's quite intimidating having two otters shuffling along and making a hissing 'ssshhh' noise, wanting to play and bite your ankles and boots.

Often we would hear a vehicle arrive, then the car door close followed by yelling and shouts, followed by running feet and some poor person flying into the office closely followed by the otters impersonating Rottweillers. It was time for them to move on and start their lives in the big dam and the riverine areas. It's not a good idea to have semi-tame otters running around the office and staff quarters, because as with all wild animals, the ones that lose their fear of humans are the ones most likely to become a problem.

My 'swimming with otters' days were cut short when they decided that my toes and fingers were a part of their dietary needs. It wasn't long after opening the cage that they moved out completely and made a small nest in the embankment and started moving down to the main dam which was less than 300 metres from the wildlife office.

On a couple of occasions when I got into the office and left the door open I was visited by these two marauders and had to shoo them out of the office, while all they wanted to do was test their teeth on my boots. Every now and then you would hear Barnard yelling from the garden, "OTTERRRRS" – and we would rush out to find him in a dance with a pair of otters, using a gardening implement to keep them at bay. Fortunately, they were very good learners and before too long they moved down to the big dam and became independent and went their separate ways.

CHAPTER 20

Spiritual Aspects

It is said that spirituality is for those seeking understanding. Any person who has lived and worked in the wild will know that there's no way that man has created what they see around them. I am comfortable in saying that a higher Being of some nature has had a hand in creating and evolving what we see in our everyday lives. Just lying on an old blanket out in the veld under the stars on a moonless Karoo evening is enough to make even the most rabid antagonist of spirituality ask: who made this and how? I know science has many explanations but certainly not all. The vastness of it all is stupendous. The Milky Way is like a road paved with diamonds running across the sky, and when it is a really dark night you can actually walk quite comfortably by the light of the stars.

Just looking at some of the geological features is enough to make one ask many questions. And yes, I understand how the mountains were created and how the continents pushed up against each other to form the most spectacular folded mountains, but it doesn't stop there. Take some time out to look at a plant, even just a leaf; the complex nature of its makeup and how it functions is an incredible testament to whoever or whatever designed and created it. I don't believe that white lions are descended from heaven as some would like to tell you. Others believe that they are the children of the sun god, Ra, and some believe that they just fell from the sky, with a bang and a big cloud of dust. This I don't give much credence to.

I also don't believe that the tail of a white lion contains a hook that can gather the stars to some point in the sky. But what I do believe is that white lions, like all lions, have a very special place in our being. They are a very important part of a whole environment and when they are not there, that environment suffers for their absence. I have a special place in my being for lions and I am spiritual. However, I am also reasonably practical and to just write off a white lion's existence to a fall from the sky or that they are descended from a sun god is to take away the amazing evolutionary journey that lions have been on for millennia.

Tales are wonderful – I love the hunter-gatherers' tales that were used to pass on knowledge to the next generation, and many references to lions are made by these gentle people. They have also left beautiful paintings and engravings of lions throughout southern Africa. According to San belief, the two pointers of the Southern Cross are lions, and there are many references to lions in the Bleek and Lloyd collection. Even in France, there are beautiful paintings of lions in some of the rock art sites. In the lives that we live there is very little time to just be, to breathe and to be quiet and listen and see what is going on around us.

I had the great honour of introducing Ian Player to the Klein Karoo and as we drove down the Tilney gorge, I could sense his whole being relax as he absorbed the beauty of what was in front of him and could visibly see the pleasure that he was getting from this incredible environment. Ian has inspired so many people to look at things in a different way, and on that day he paid me a very special compliment by telling the people at dinner that night that I was one of the few people who know when to just be quiet.

In nature, if an animal didn't stop, look and listen to their environment it wouldn't live for very long. Lions are respected and loved by those who have read about them or who have had the privilege of observing them. They are reviled and feared by those who haven't observed them or done any study of them. They create so many different reactions in different people and I've been privileged to observe many of these. They range from absolute terror to the most incredible awe and respect. The hunter-gatherer people, sometimes known as the San or Bushmen of southern Africa, have left their exquisite paintings of lions and other animals in rock shelters and overhangs in thousands of places around the country.

In places where there are no caves and overhangs they made engravings on boulders and flat rocks. These engravings are known as Petroglyphs and are found mostly in the dry central interior of South Africa. The paintings and engravings are so perfect in all aspects that they are truly Africa's art treasure. Some of them date back 27,000 years, so they are also some of the oldest art forms in the world. The paintings and engravings were created after a trance dance when the Shaman or members of the group would do a healing dance which resulted in a trance state. The person who entered the trance was, in their terms, actually dying, and when asked if they were truly dead, the answer was yes but they would come back soon. The person then visited the spirit world while in the trance and communicated with the spirits for guidance and healing and sometimes even requested rain.

It was after the trance that the drawings and engravings were lovingly painted or chipped. In the western Cape and the Drakensberg the most commonly painted animal was the eland. The eland is the largest antelope species in the world and the bulls are massive, weighing up to 900 kilograms and standing up to 183 centimetres at the shoulder. Yet as big as they are, they are incredibly nimble and can jump fences with consummate ease and can cover vast distances very quickly. They have an easy, trotting-type of rhythm that is deceptively slow-looking but when they get into their stride they can move unbelievable distances.

The hunter-gatherer people had extreme respect for the eland; its tenacity and endurance were legendary and the meat and fat were highly prized for their sustenance. The skin of an eland was highly valued as a kaross and for making bags and other items of clothing. The eland was the animal that their young men would have to hunt for their rite of passage into adulthood. When an eland was killed they would gather around the animal to absorb its power, and when they used the fat and blood to create paint for their art, it was as though they were putting the power of the eland into the rock.

I have done some pretty crazy things in my life but this one is right up there at the top of the list. We had done a very successful cheetah release and were enjoying Ryno's culinary skills around a fire with a braai. Ryno makes the best 'braaied' chicken in the world and although I'm not exactly sure of his recipe, I do know that copious quantities of a product called Aromat and some lemon juice are part of it and I have yet to taste better chicken. After a couple of drinks, Paul came up with the idea that we should smoke some elephant dung. His theory was that at a specific time of the year the ellies would eat some plants that were hallucinogenic and that if we smoked it, we would get into a trance-like state and be able to see what the bushmen saw.

It is amazing what influence one too many glasses of red wine has over common sense. On a beautiful evening, in front of an open fire we sat around and smoked dry elephant dung. I'd given up smoking tobacco products ten years earlier so my lungs were relatively sensitive and I coughed quite badly but persevered in the name of science. It wasn't long before the effect kicked in and I was walking on hot coals and doing other stupid stuff. I went to bed and to this day have no recall how I got there, but the next day was not an easy time for any of us. All I can tell you is that two truths came from this – one is that you definitely hallucinate when you smoke elephant dung, specifically when the elephants eat plants with hallucinogenic properties, and secondly, if I was put in a cave with some paint there would be absolutely no way that I could create anything that resembles the beautiful rock art that the San people created. And the last thing is – I will never, ever attempt this again!

CHAPTER 21

GRAVES AND GHOSTS

After the First World War many areas around the world were affected by what was called the Spanish Flu. It was the single most devastating episode in South Africa's demographic history and in the months of September and October of 1918 alone, more than 140,000 people died in South Africa. The final death toll from the epidemic would exceed half a million people. Worldwide, the Spanish flu pandemic started in January 1918 and went around the world, even to the Arctic and some of the most remote islands. Globally, it was estimated that over fifty million people died and over 500 million people contracted it before it was brought under control in December 1920.

I was examining a specific area on the reserve where the vegetation had been historically disturbed when I started noticing rocks that had been packed into definable rows. When you fly over parts of rural South Africa you will see small circular patterns that occur in a haphazard way, chequered across the landscape. You will also notice that the vegetation on these circles is different from the vegetation surrounding it. These are known as 'heuweltjies' or earth mounds and some people call them Memo mounds. From the ground they are clearly defined mounds of soil varying in size from about 10 to 25 metres in diameter and they can get up to about two metres tall. These mounds are created by the activity of the harvester termite Microhodotermes viator and carbon dating has indicated that some of them are up to 4,000 years old. The vegetation that grows on them is quite distinct from the surrounding area, making the mounds clearly visible to a person who knows what they're looking at.

In a specific area quite close to Tilney Lodge, I found the first set of graves and it was quite obvious why the mounds would be used as burial sites. The soil on the mounds is soft and porous created by hundreds of years of termite activity, so compared with the normally rock-hard soils of the Klein Karoo the mound's sandy surface is much easier to dig into. The graves are marked out by a series of packed rocks and they indicate the length of the grave hence also the possible age group of the occupant. Some have a headstone in the form of a boulder standing higher than the rest, but I haven't seen any etching into these rocks to indicate who the occupant was. Some even have a head and footstone that protrude higher than the rest of the grave markers. I'm almost sure that the rocks were packed in a specific way to stop scavengers from digging up the remains of the person; however, I am also of the opinion that the rocks were put there to demarcate the area as a grave.

While doing vegetation observations around the Tilney area I came across many of these mounds with graves on them, and I started to do some research on them. My research started in Montagu and it was incredible to read the devastation caused by the flu epidemic. The people of Montagu didn't suffer as much as most other places in the western Cape, however, because the district surgeon had cleverly closed off the village. People were free to leave but nobody could enter again till the epidemic ended in 1920. Luckily for everybody, Montagu wasn't difficult to protect as there is a most beautiful kloof that you have to go through to enter from the south west. Cogmanskloof, as it is known, is in the Langeberg mountains with steep sides and a tunnel that was created in the 1800s, and in 1899 during the Boer war the English built a small fort on top of it. The eastern and northern sides were more difficult to protect but they did a good job of it as not one person contracted the illness.

I've been told by a couple of the older citizens of Montagu and Barrydale that the graves at Tilney are from the time of the epidemic. Before the lodge was built at Tilney there was only the manor house, which now forms part of the lodge. We sometimes used it to house people visiting us before the new facilities were built. Quite a few of them told me that they had heard very strange noises in the night and some even described a ghost-like figure moving around. I have slept out there quite a few times but have never had a problem with any spectral apparitions. Interestingly, over the years, I have been told by numerous people that there are certain places on the reserve that are populated by spirits. Many of these people don't even know each other, yet point out the same places. After this had happened a few times from very different people, I came to the conclusion that there must be something in those environments that I didn't quite comprehend.

With spectral apparitions in mind, it is truly amazing how quiet an elephant can be. Before the fence was erected around the new wildlife offices and staff living area, we used to get quite a few nocturnal visitors. One night I was awakened by a strange noise outside my bedroom window and reached over, picked up my torch and lit up the area around the open window. It was very strange as all I could see was grey, and only when the grey started moving did I realise that it was an ellie right there next to the window, less than a metre from me – I could have reached out and touched it, it was that close. What was wonderful to experience was just how quiet these great big animals can be. As soon as I switched on the torch everything went quiet and only after they had relaxed and started eating up my garden again could I hear them.

I was once at a friend's reserve watching a group of five elephants rolling and playing in a small mud hole. There was the matriarch and the rest were made up of teenagers and one two-year-old. They were all enjoying themselves when the female got up and started ambling down the road leaving the youngsters to play. When she got about 70 metres away she made a very low 'brrrrrr' sound to which the four youngsters reacted immediately and ran to catch up with her. There's no doubt that it was an instruction which was obeyed instantly to the point that the youngsters were scrabbling over each other to get out of the mud and onto the road.

CHAPTER 22

LIONS AT WAR

One of the most testing moments of my life occurred when the southern pride was released from their boma. They stayed down in the south for a while and became legendary hunters. I witnessed the three of them kill two gemsbok within metres of each other. The young male headed north and we were all of the opinion that because the northern pride's male was far bigger, he would be pushed back south very quickly – but how wrong we were! This youngster broke up the northern pride and started pushing the females further north. The northern pride then disbanded and dispersed throughout the northern sector.

Just when it looked as if peace was being restored, we got an urgent call from the fencing crew that a lion had gone through the fence to the north west. This area was still in the reserve's property but the lion was now outside of the protected area and there were no game-proof fences between the lion and our neighbours. These farmers are livestock farmers and they rely on their cattle and sheep to survive.

They are good people indeed and I had, over the years, made a point of visiting them and keeping them up-to-date with what we were doing on the reserve. Some we had problems with, but it was normally just unfounded fears that we were becoming a 'problem animal breeding centre' as they called us, and that all of the jackals and caracals that gave them problems on their farms originated with us.

Caracals are sometimes incorrectly called 'lynxes,' which are from Europe and very different in colour and size. The name caracal is possibly derived from Turkish Garah-Gulak which translates to black ear, their descriptive Afrikaans name being 'rooikat' which means red cat.

At the start I'd invited all of our neighbours to the reserve so that we could show them what we were doing and what our future plans were. It was a very productive day and we were able to change some of the negative attitudes towards us. It was always a good idea to invite their wives along as well, as their calming influence made it easier to communicate. We took them around the reserve and then at the wild life office we showed them our research projects, what we were doing as far as the management of the reserve went, and also what measures we would take in the event of an escape. Over the years I had built up a measure of trust and it was important to keep that intact, but a free-roaming lion in a domestic livestock area will always go for the easy meal, so we had to act swiftly.

I called the team together and briefed them on the situation and asked for input. We then decided that we would go out and attempt to dart the lion; only if we didn't succeed on the first attempt would we advise the neighbours and authorities of the situation. I sent the team off to get prepared and to collect the necessary equipment. In those days the reserve didn't have any firearms and the dart gun was a rather decrepit 'hand me down' from Shamwari. Ryno and I had our own weapons for culling and boma feeding of the predators. In conservation, firearms are not only a reality but a necessary tool of the trade. I picked up my rifle, a .308 calibre and 44 magnum revolver, both capable of stopping a lion but not the ideal in a charge situation.

The idea was that Jan would carry the dart gun, Ryno his rifle and I would carry mine and we would approach the animal on foot if we couldn't get to her by vehicle. We would use the students and Paul and Karlien as our eyes and put them on top of a ridge that looked down on the dry river line where we suspected she was. Just before we were about to head out, Ryno told me that he was uncomfortable with the plan and wanted out. He is a good marksman and I really wanted him as a backup. However, I knew what we were going into and understood his trepidation and asked him to go up onto the ridge as a stopper and an extra pair of eyes if she ran away from us. I then had Jan with the dart gun, Paul with my 44 Magnum and myself with a medium-calibre rifle. Before we went out I removed the telescopic sight and the scope mount, to facilitate very fast shooting in the event of a crisis. The briefing went something along these lines – we were going to attempt to get within range to dart her from a vehicle, but in the event of that failing, we would go in on foot.

If we could manoeuvre a situation where Jan could take a shot with the dart gun, he would do so. I would protect him at all times and in the event of a charge he was to fall back and Paul and I would evaluate the situation. The tension was tangible as we attempted every avenue possible to get close to her in the vehicle, but to no avail. The decision was made for us and I briefed the team again and made sure that Ryno was in the cut-off position on the hill and that Karlien and the students were in a safe place. We searched for the lioness everywhere along the roads and eventually found her spoor heading into the river line, which was the worst case situation as we would now have to go in on foot.

The acacia trees on the river line have long, sharp thorns and because the area retains more moisture, the vegetation is thicker than in the surrounding area. We moved into the thicket and crossed the river bed to get to the high ground which I hoped would give us a better perspective and a bit of height to be able to see her. As we crossed, she was just to the south of us and gave us a warning growl which vibrated through every inch of our beings. We couldn't see her and started to climb up the ridge when a flash of tail and a deep growl to the right gave away her position. We stood dead still and waited. She was in a thicket slightly below and about twenty five metres away, a perfect distance for the dart gun, but she was in the thicket so Jan couldn't take a shot.

I was hoping she would relax and give us an opportunity to dart her. Our nerves on edge already, we were suddenly shaken when out of nowhere, she charged. It was so fast that Jan didn't have time to get a dart off. Just as quickly as she'd charged, she was back in the thicket under cover again. Feeling jittery and anxious, we waited some more, standing motionless for what felt like hours. I finally decided to push forward, trying to intimidate her, and found myself in a crisis situation. She charged again but this wasn't a mock-charge – this was the real thing! She was now intent on attacking us and charged at lightning speed. I can't describe the noise she made, other than that it was truly terrifying, the swiftness of the attack unbelievable. I've been mock-charged by lions on quite a few occasions over the years and knew what to expect, but the speed at which it happened was bewildering. I had two people with me who were in mortal danger and I had only split seconds to make the horrible decision to stop her.

The explosion of the shot was followed by a terrible silence, even the cicadas went quiet. I couldn't see her as she'd fallen into a bush just in front of us, and the recoil of the rifle rising upwards had prevented me from seeing the result of the shot. I needed to know that it was over. As the shot went off I reloaded, without taking my eyes off the area where I'd fired. Jan had got behind us as the dart gun had become superfluous and Paul and I approached the area very slowly, ready for anything. Every nerve in my body was at breaking point, and even though I didn't take my eyes off the area where she was, I was hyper-alert to everything around me. I couldn't see Paul but I knew that he was to my right and slightly behind, exactly where I'd asked him to be so that I wouldn't have to worry about him if I had to shoot again. It took what felt like an eternity to cross those few metres but when we saw her she was still. The shot had penetrated her brain and she had dropped in mid-air. What started as relief, soon turned into dreadful emotions of guilt and sadness. I've asked myself every day since then if I could have done anything differently to change the outcome; but try as I may, I can't shake the hollowness I feel every time I think of that beautiful lioness.

The students helped with the removal of her body and I debriefed the team. It was a very sad day for all of us, as unlike most of the other lions, she had been given a name. She was also one of the first lions introduced to the reserve and also the first one to give birth, albeit unsuccessfully, in the Klein Karoo for centuries. Due to her sexual prowess she had been nicknamed Pompie, a rather risqué Afrikaans word for the act of procreation. Many a time when the wildlife team were monitoring the lions you would hear over the radio that Pompie was in a specific area doing whatever it is that lions do. I was rather glad that the wildlife team and the rangers had different radio frequencies as I felt that if the guests had heard what went over the airwaves there would have been a lot of explaining to do.

When I got back to the office I had the unenviable task of phoning the bosses and fortunately for me, when I phoned Adrian Gardiner to give him the sad news, he was very gracious and supported my actions. I was truly appreciative of this, as I'd been expecting a tongue-lashing, which I felt I deserved. He was very encouraging and understood why it had had to be done and for this gesture I'll always be grateful. I know that this incident affected the whole team and the students in different ways and it was important for me that everybody understood why it had happened and what we could do to prevent a recurrence. I sat quietly to one side while Karlien and Paul explained to the students the direness of the situation and the reasons for what had happened. The two of them are incredible teachers without actually being qualified. The questions were answered candidly and with full explanations and they held nothing back, and the students were satisfied with the answers.

Ryno and I had a chat afterwards and I appreciated his honesty and the fact that he stood up to me, knowing that his skills as a marksman in keeping the students safe in the event of a problem, made my work that day much easier. For Jan and Paul's bravery during this incident I can't ever say enough. Plus I can't imagine what Jan must have thought when I'd asked him to walk in with me on a stressed lion, carrying only a dart gun; and Paul must have thought that I'd lost it completely when I asked him to back us up with a revolver. They didn't flinch – they just joined me and did exactly what they'd been asked to do. My whole team gave their unconditional approval of my actions during this sad event and for this I'll always be grateful. However, it doesn't take away the deep sadness.

The team did an autopsy on the lioness; she was in excellent condition, showing no signs of any disease or sickness. When I looked at her skinned body and had the opportunity to examine the muscular make-up of a lion up close, it reaffirmed everything I knew about these felines. They are incredibly well proportioned for their function as a predator of medium to large antelope. The muscle groups are structured in such a way as to complement the next group and slot in perfectly with each other. They are so detailed and precise that when you look at each muscle you can see exactly what it's used for. I was reluctant to tell anybody outside of the reserve of the incident, but this is the Klein Karoo and things have a way of getting around and the stories also have a way of turning into fantasy and taking on mythical proportions. So I reported it to our local conservation officer, and the next day decided that instead of phoning the neighbours it would be better to tell them face to face.

It was a very interesting morning and it was the right decision because the embroidered stories were already making the rounds and I was able to tell our side before it got out of hand. This, however, didn't stop the rumour mongering; I was in the farmers' Co-op in Barrydale later that week, listening to two farmers tell of how a whole pride of lions were running around eating sheep and anything else that got in their way. It took quite a lot of composure on my side not to laugh or react to the stories, and even after a farmers' meeting where I had explained what had happened for the hundredth time, the stories still persisted.

CHAPTER 23

ELEPHANT HAPPENINGS

We had introduced a small herd of elephants to the reserve, and were really blessed with the animals that we received as they were calm and relaxed around vehicles. The twenty-plus-year-old bull acted far more maturely than his age; he was a pleasure to be around and he gave many people unforgettable memories. An elephant normally starts showing signs of 'musth' at the age of twenty four, although they don't complete a full three-month cycle until they reach thirty five. Our bull started small 'musth' cycles but luckily was never aggressive towards people or vehicles which they can be prone to do.

I have a special place in my heart for elephants. As a young soldier I used to climb into the trees on the Caprivi Strip and count the herds coming across the border from Angola and Zambia. Some days when I was up there the whole area around me was a sea of grey, hundreds of elephants surrounding me in the tree, going about their daily lives. Elephants of every size and age would move through this area on their way down to the Okavango swamps in Botswana. I regularly counted herds of 500 and more; sometimes I would be stuck up the tree till nightfall until the last of the herd moved away.

It was also here that I learnt from a Bushman/San tracker how to approach elephants on foot and all of the signs to look out for. There was a rather cantankerous old bull living in the area that had damaged one of his tusks on a tree trunk and I'm sure it was this exposed nerve ending that made him so unpredictable. Moffat, the tracker who was basically assigned to me to protect me from myself, and I, would often go and have a good look at the old bull if the wind was right and he couldn't hear us. From this man I learned so much, even though his command of the English language was zero and my understanding of the San language was less than zero. However, we got on extremely well and language didn't seem to affect our understanding of each other and what we wanted to convey. We were acclimatising a new batch of troops to the area, which involved showing them a specific area and what to expect when on patrol. The leader approached and said that there was an elephant in front of us. I answered, "Don't worry, just walk forward and the elephant will move out of the way."

Off he went and literally ten seconds later came back and, in the most poetic Afrikaans, told me that I could go and take a flippin' hike and walk at the flippin' elephant myself because this flippin' elephant's ears were flapping backwards and flippin' forwards and he flippin' wanted to stomp on them. I could see that he thought I was totally insane when I explained to him that this was their method of cooling themselves and if he just walked at the ellie it would calmly walk away.

His answer was something along the lines of, "Not a fuck am I going to walk calmly at that fucking elephant. You're fucking bush mad enough so you can fucking do it." I got slowly to my feet, went around the bush and my heart stopped as there stood our grumpy old bull, literally slap bang in front of me. Bulls over the age of thirty are really very impressive and can be as tall as four metres; and added to that the fact that he was lifting his head and flapping his ears which made him look taller, I wasn't surprised the young lieutenant wasn't going to walk at the elephant. Neither was I because as I got there, he didn't even hesitate, he just charged. The noise was terrific and I can't even begin to describe it. We all scrambled in different directions, swearing, howling and screaming, kit flying everywhere. I actually saw one chap run straight through a 'wait a bit' bush which is full of hooked thorns that, under normal conditions, arrest your movement instantly and completely. After the dust had settled we found a chap up a tree which didn't have one branch or any form of grip on it for the first five metres! We had to make up a rope system to get him down.

Elephants, it turns out, are magical and capable of giving humans the power to do miraculous things. Fortunately for us, that ellie just went straight on ahead without turning back and that was the first and last full charge by an elephant that I've ever experienced. I don't think that it was a fury charge at all but it certainly got my full attention and the young lieutenant was now more than ever convinced that this bush-mad, elephant-loving 'engelsman' was more of a danger than the flippin' enemy.

Over the years I've grown to love elephants and have spent many wonderful times in the field monitoring them and just absorbing their behaviour. The area below the wildlife office is a flood plain and a favourite spot for the small herd, so in the late afternoon I would often go and sit on the ridge and just watch the ellies 'moseying' around. It is in these moments that I really am at peace with the world and myself. I'm not sure what it is about them that makes me react that way – perhaps it's their size and gentle attitude towards each other. They are incredibly good-natured, always touching one another with their trunks and bodies. It's almost as if they reassure one another that everything's okay and if anything alarms them, the females gather the youngsters under them and form a barrier with their bodies till the danger is past and then they reassure the youngsters by tender touching and soft noises.

There was a cloudburst one afternoon and Jan and I went down to check the fence where the water enters the reserve. The ellies were on the far bank looking at the raging torrent going past when one of the year-old, pocket-sized youngsters fell in. We stood there, appalled at the thought that we were going to witness this baby elephant drown. The youngster hit the water and the current took it downstream and we jumped into the vehicle and drove along as close to the flood waters as we could, every now and then seeing his little trunk breaking the surface of the bright red silt-laden water. I had ideas of following it till the water entered the dam and then diving in to see if we could rescue the little chap.

What then happened was a real eye-opener. Mother ellie calmly walked into the raging water and trotted after baby. She swiftly caught up with the youngster and gently lifted it out of the river and up the bank on our side. She then climbed out herself and comforted the 'pocket size' which didn't look any the worse for wear. In fact, it looked decidedly content with life and its adventures.

My father, Jeff, wasn't really a bush person till late in life when I tried to give him the best experiences I could arrange. I took him to see the elephants one afternoon, and they really gave him a wonderful show. The bull walked straight over to the vehicle and stopped literally a metre from his door and just looked at him. I could hear Dad suck in his breath and told him to just relax and watch. The bull was very even-tempered and started eating and then slowly moved his way around the front of the vehicle. Then two of the females with their calves did exactly the same thing. He was awestruck and when we got back he couldn't stop talking about his experience. Karlien picked up on this and named the bull Jeff; it was a great honour that sadly he was never to know about.

When Dad was in the last few weeks of his life's journey with us, Simmy would phone me every day to keep me updated on his wellbeing. I spent as much time as I could with him and, in fact, we had some very special times during that important period. He never gave up and the last time that I ever saw him, his computer was giving some problems which I fixed and he then asked me to do the replies to his emails and to send out his newsletter. Such was his character that a week before he passed on, he was still telling his cousins in Canada that he would be visiting them soon.

I suppose that I knew his time was coming to an end but it was just happening a lot faster than it was supposed to. Personally, I think it was all the drugs that were being pumped into him.

I briefed the team and they knew that I would have to leave the reserve for a while in the not too distant future. Understandably, I wasn't functioning at 100 per cent capacity, so every evening after my inspection tour I would end up sitting on the hill looking at the elephants. They gave me a lot of comfort and later that same week when I got back to the office and the phone was ringing, it was Simmy sadly telling me that he had passed on. I went straight back down and spent till darkness with the ellies.

A few months after he passed away, a very strange incident took place when my mother visited us on the reserve. I've never told my Mum that the bull was named Jeff; she will only find out when she reads this book. Mum was in the passenger seat right next to me and I parked about fifty odd metres from the herd when, first, the females walked over and stood on my side of the vehicle looking in at us. I could sense my mother's unease and told her to relax and enjoy what she was seeing. They then moved around the vehicle and looked in at her and just stood there looking at her for about thirty seconds. I kept telling her that they had come to greet her and she relaxed a bit. And then the bull came over to the vehicle. He approached right up to the vehicle, his trunk literally at the door. He looked in on me then moved right round and came to a stop at my mother's window. I couldn't believe what I was seeing – there was this massive bull elephant looking in at Mum. He just stood there for at least a minute and it was as if he knew exactly who she was. Simmy was in the back seat and when I glanced in the rear-view mirror I could see exactly what she was thinking – and she could also read my thoughts. We were both nearly bursting with emotion. When he finally walked off, I had to look away to hide my feelings.

CHAPTER 24

QUEEN

For me, Christmas is a time for families and it has always been important to me that my team get time off over this period to be with their loved ones. The same cannot be said for New Year's Eve as normally the team would arrange a party of epic proportions at a venue close to the reserve, to be with all of their friends. Due to distances, I didn't regularly get a newspaper on the reserve and relied on the radio to keep me updated on what was going on in the world around me. On my early morning tour I would listen to the news and weather on the different channels.

On the morning of 26 December I was on my early morning inspection (it was just after four in the morning) and I was driving over the rise next to the dam on my way down to the lions. On the news there were reports of a colossal catastrophe of some nature starting to unfold. At five o'clock the news readers were talking about a loss of life of more than five thousand and literally every hour that figure rose to a point where, later in the day, they were talking in hundreds of thousands. A tsunami in the Indian Ocean had struck Indonesia and many other countries, causing total devastation and huge loss of life.

When I got to the lions at about five thirty they were lying down next to a river line with a thicket behind them which gave me an excellent opportunity to take my time and have a good look at them through the binoculars. Jabulani and Queen and two of the cubs were in pretty good shape but one of the cubs was still full of ticks. Soon after their release we'd had good rains which often bring out the parasites. Their faces got covered with ticks that were swollen with blood and looked like grey blotches which did nothing for their cute appearance, but more importantly, they didn't seem to have any adverse effects physically. The ticks are more obviously visible on the white lions than their tawny counterparts and they really don't look very good when covered with great big fat blobs.

I had often noticed ticks, mostly on Jabulani and Queen but now they could be seen on the youngsters too. We kept a very close watch on the situation because at one stage they really didn't look good at all, full of blood engorged ticks, mostly around the eyes and cheeks. For the months of January and February the pride really looked in bad condition, but as soon as March arrived it was as if something had come in and cleaned them up. I have a suspicion that some of the prey species had ingested aloes, and when the lions had eaten the prey, the effect of the residue of the aloe in its bloodstream had caused the ticks to drop off the lions.

The reason I say this is because I had been watching some kudu in the hills eating aloes as though it was the last food source on the planet. I was surprised at this because if you have ever had even a drop of aloe on your tongue you'll know what I'm talking about. It's got to be one of the most unpleasant taste experiences I've ever had, and to watch the kudu gobbling it down to the point where they had the juice dripping down their jowls, makes my mouth salivate. I'm now convinced that they take in the aloes to help keep the parasites in check; and when they're eaten, they pass on this natural insecticide to whatever preys on them.

I was relieved that the rest of the family had lost a lot of the ticks and it was while I was having a close inspection of them through the binoculars that I noticed that Queen kept looking slightly to her left, back at the thicket. I glassed the area but couldn't see a thing. She kept glancing over and then, as quick as a blink, she jumped up and charged into the thicket. I couldn't believe my eyes, as, right there in front of me, she had taken an adult male duiker. The duiker is a small antelope that gets its name from their habit of running at high speed and jumping or diving into cover. 'Duik' is the Afrikaans word for dive. Only the males have short horns, and they weigh in at about 18 kilograms. I watched in absolute amazement as she killed it as if she'd been hunting her whole life. It didn't take long for Jabulani and the cubs to join in and the duiker was very quickly devoured. On my drive back to the office I was one very happy person and was looking forward to telling the team and the bosses. But the team was away on Christmas break, so they would only hear about it on their return; but I didn't take long in letting the bosses know. This was tremendous news and it was always a relief for me to give them good news of any sort.

December 26 will be in my memory for two very different reasons – the tragedy of the tsunami and the day that Queen made her first kill. Although we had never factored in the possibility that Queen and Jabulani would hunt, it was a very big step for the cubs to witness their parents hunt and consume what they'd caught. Up to then they knew that their parents supplied the food but they'd never seen how they did it. Now that we knew that Queen was capable of hunting, we would monitor her behaviour closely to see if she was capable of supporting her family. We wouldn't stop feeding them unless they became so successful that it was no longer necessary; however, the chances of that were rather remote.

One female and four males wasn't exactly the ideal balance at all. Although young males are capable of hunting, they don't do well and most survive on scavenging and hunting smaller animals. Only ten per cent of male lions make it to maturity and even fewer get the opportunity to become a pride male, so once the pride was out in the larger area our method of feeding had to be carefully planned. The idea was that they would wander onto the carcass after we had dropped it off. The biggest problem to overcome was not letting the cubs see a vehicle drop off any food whatsoever. It was critical that they didn't associate humans or vehicles with feeding.

As the youngsters got bigger the competition at feeding started to heat up. If the carcass that was being fed on was small like a springbok, Jabulani would often pick up the whole thing and take it off for himself. It was sometimes a tug of war with Jabulani on one end and three cubs on the other, with Jabulani always winning the day. Queen would often wait in the background and move in on the food when the other four were slowing down with extended tummies.

There were two entrances into the area and before any of the team doing the feeding even got close, they would stop and use the telemetry set to ascertain where the pride was and then go in at the furthest gate away from them and find a suitable shady spot to drop off their food, which was always in the form of an entire carcass. When the animals had been in the habituation phase in the boma, we would often put a dietary powder supplement into the cavity of the carcass we were feeding them. This is a stressful time for any animal and the extra nutrients and minerals help get them through this phase. On a fairly regular basis we continued with this approach after the white lions were released. I don't think that it was totally necessary but it certainly didn't do them any harm either.

CHAPTER 25

THE RESERVE PRIDE

The reserve pride had settled in nicely and it seemed that a sense of normality had finally come about. The two females had both reached sexual maturity and lots of mating was on the go. A typical pride situation had settled in with the male protecting the females and the females doing the hunting. It was a very interesting project, monitoring their movements. For about a month all of them were together and then, out of the blue, the male would head off by himself for two or three days, literally covering the whole reserve and marking his territory. This would be accompanied by a series of roars that shook the air, warning any other male lions that this was his turf and they would meet with opposition should they try to move in on him. After a few days he joined the females again and this reunion was always delightful to witness. Lots of neck rubbing, head contact and tail flicking, one of the most expressive rituals that lions carry out when they reunite, and always very special to watch.

A female doesn't display any signs of being pregnant until the last week of pregnancy; however, because of our monitoring procedures we knew that both of them were due to give birth in the next week or so. So we stepped up the monitoring process to observe what was going to happen. One female started showing swollen nipples and this is a sure sign that there are only a few days left before birthing. After a few days we were all a bit nonplussed by her behaviour. Normal behaviour would have been that she would leave the pride and find a thicket or a cave to give birth in. We then found the reason why – she had given birth to a single cub right out in the open on the flood plain. Shades of what had happened before with Queen … This didn't look good at all and the chances that the youngster would survive were not very high.

In the afternoon I went down to see if there was any change in the situation. The male was standing next to the little cub and nudged it with his paw. It rolled over but did little else, whereupon he then did something that I caught on film that will always raise a question of the sensitivity of animals towards one another. He stood next to that diminutive little creature and roared for a good few minutes, which brought the female out of the thicket. As soon as she arrived he walked away and not long after, the mother also moved off. To make matters even more tense, a herd of elephants moved towards the area where the youngster lay. They scented the area and could obviously smell the lions and the cub but they didn't see it. A large cow got so close that she nearly stood on the cub – it was literally centimetres away from being crushed. Fortunately, after inspecting the area the ellies moved off which then gave me the opportunity of getting in a little closer to inspect the cub. It was hardly breathing and had no chance of surviving.

This is one of the hardest decisions that any wildlife manager must make – do you take the dying cub and attempt to hand-rear it or do you let nature take its course? This is when all of the 'what if' and 'if only' questions run around in your head. It's never a cut and dried decision; it depends on the merits of each case. I have the utmost sympathy for anybody who has ever had to make the choice because it's one of those situations where you are damned if you do and damned if you don't. There will always be some who will agree with your choice and conversely, there'll always be those who condemn you for your decision. Mostly we have to live with ourselves for our decisions, which is sometimes harder than any form of criticism that your peers can mete out.

That afternoon I had about six students with me and after having a chat with Ryno, my decision was to let nature take its course. This created a very lively debate amongst the students, which we have always encouraged, knowing that there is not one correct answer to any dilemma. Every part of any nature conservator wants to protect the animals that they love so much, but sometimes love is tough and the idea of bringing up a lion which would then face a life either in a boma or behind bars, is not what any conservator wants either.

Later that afternoon I decided to go back and see what was happening, and also to take some photos and a video to show the rest of the team. The decision was taken out of my hands because as I was taking the footage, the little animal stopped breathing. This cub was the lioness's first birthing and not all mothers are perfect first time around. As far as I could tell the female didn't come back to the area and I suppose it was her way of dealing with the situation, to just get on with her life. When we replayed the footage it quickly became apparent that the little one would not have survived even if we had brought it in to try and hand-rear it. The male returned just after the little one had expired and stood over the small body and roared for at least five minutes. He lay down next to it for a while, then got up and headed east to join the females.

One afternoon this same male, who had always had a very relaxed attitude to life, gave me a very big wake-up call. I had a group of students with me and was busy doing some lion monitoring in an area with lots of acacia thorn trees. I had to stop to tell the folk on the back of the vehicle to duck down when we went underneath the trees. We found the male underneath a tree, very relaxed and resting in the shade. After having a good look at his condition I reversed back and was in a situation where I had to put the back of the vehicle under an overhanging branch. Kelly, a young Australian student, was in the front with me and I told her to tell the gang in the back to duck. She did this and I began to reverse. The next second I heard a scream and stopped instantly. The lion, who was about thirty metres away, had stood up to see what the noise was; I told everybody to hold on and drove about 100 metres away from the lion, stopped, jumped out and in the back was Suzy who looked as if she was bleeding to death. I grabbed the first aid kit, told the rest to watch out for the lion and inspected Suzy's head. There was a small cut at the top of her forehead just inside her hairline which, as all head wounds do, was bleeding profusely. The blood trickled down her face creating a very scary impression indeed.

I'd just begun to clean the wound when Kelly called me and said that the lion was approaching the vehicle. There he was, padding his way towards us. I got into the vehicle fast but I needed to put a compression bandage on Suzy's wound, so told everybody to sit still and keep quiet so that I could observe what the lion was going to do. It was amazing; he headed straight for the vehicle and particularly for the back of it. I slapped the door which stopped him in his tracks, and then sped away for about a kilometre, stopping to fix up Suzy's wound properly. I'm not sure if it was relief, but Kelly started giggling which got the whole lot going again and then ... there he was again! The lion had followed the vehicle and didn't even hesitate but came directly towards us. I wanted to see what he was up to and drove in a large circle so that I could watch his reactions. I was absolutely amazed as he followed the vehicle, concentrating totally on the back where Suzy and the rest of the students were sitting. I decided to get back to Gondwana and patch up Suzy properly. I've never seen this type of behaviour before, and am still not sure what got the lion going – Suzy's scream or the smell of her blood. Fortunately, he never displayed this behaviour again.

CHAPTER 26

FUN AND GAMES

The white lion project was nearing a very important phase as the cubs were now heading towards becoming sub-adults and, in the wild, the young males are normally ousted from the pride at this stage. This can occur at any time from the age of about 20 to 24 months. Queen and Jabulani were playing their parental part to perfection and had raised three balanced cubs. There was one specific cub which started approaching the vehicles, not in an aggressive manner, but out of curiosity. It became obvious that no matter what we did, he would approach too close. He would even get right up to vehicles that were never utilised for the dropping off of food. It's never a good idea to have guests in an area where a lion has no fear whatsoever of vehicles as it can lead to a range of unpredictable reactions from a lion if someone on the vehicle makes any sudden movement.

Hopefully the lion would back off but one can never be 100 per cent certain that it wouldn't trigger a reactive response ending in a charge, or even worse, a full-on attack.

With the lion being so close nobody would have time to react quickly enough to protect the people on board an open game-viewing vehicle. With this in mind all of the staff were warned to keep a reasonable distance and if approached, they were to evaluate the situation and, if necessary, drive away from the lions.

John and Tina Simpson are a husband and wife team of film makers. John had been a military man for most of his life, and this military influence would often be portrayed in his mannerisms and his way of dealing with people. I soon found out that underneath his soldierly veneer he was actually a very soft and gentle person who loved his family and really enjoyed the wild places in Africa.

I'd been asked to take him out to do some filming and late one afternoon we were about 70 metres away from the lions on their western side. It was a beautiful afternoon with the orange glow of the sun casting shadows on the lions, absolutely perfect for filming and John got busy setting up his equipment. He asked me to move the vehicle backwards as the shadow of the vehicle was getting into his frame. He released and pulled in the camera but left the tripod outside. I advised him that it wasn't a good idea and that he should bring the tripod in through the open window. John, being John, said, "Go backwards and hurry up as I'm losing light. We can pick it up later." Well, I reversed about twenty metres and stopped, John resting the camera on the vehicle's window as a stable platform and continuing to film. In that special sunset glow the lions looked incredible.

For a while all went very well until Jabulani started to stretch and yawn – a sure sign that he was going to get up. I told John that we should pick up the tripod before he got going. "Don't worry about it," was the answer. Well, Jabulani did get up and very slowly made his way towards the tripod. By this time it was too late for me to do anything and, of course, by this time John was telling me to go and get it! I've never seen a lion behave this way and it was hysterically funny.

He approached the tripod with absolute trepidation, putting one paw out and trying to reach it but just as he got near he would jump back. He touched it once and the top moved which startled him and he rushed away. He slowly returned and this time got so close that he started sniffing it, then biting the rubber strap that held the legs together, but by this time John was busy filming the incident. Jabulani then bit down on the leg and ran off with the tripod between his teeth. He ran right past Queen and scattered the cubs in all directions, then moved off to lie down close to the thicket. By this time the cubs had joined in and it was a right royal tug-of-war with the cubs on one end and Jabulani on the other. John didn't see the funny side of this at all, and kept asking me when I was going to do something. The more I explained that I couldn't do anything about his tripod till the pride moved off, the more he kept going at me.

It took quite a while for them to get bored with the tripod and move off, by which time, fortunately, John had regained his sense of humour and commented on the fact that this would be one of the most unusual insurance claims, but that he had film to corroborate his story. We eventually got the opportunity to retrieve the tripod and, taking into account the amount of time the lions had been playing with it, it was (surprisingly) still in fairly good condition. The rubber spreaders were destroyed and one leg had a few puncture wounds. When John tried it out he could still use it, with a few running repairs.

We later went on to collaborate on a film and the footage of that incident is in that film. Tina and I spent many days behind an editing suite in her studio getting the film ready, she doing the editing and I making sure that we had all of the animals and plants named correctly. I had no idea at that point just how difficult it is to edit a film. Tina is a very talented person and her editing skills are incredible, with a great sensitivity for what she does. It was also at this point that I came to the conclusion that it might just be that I had it in me to make a film. This was to take me down some incredibly interesting roads over the years to follow.

CHAPTER 27

Students and Whales

When the first students to stay on the reserve arrived, they were in small groups, sometimes as few as six. And to give my team a well-deserved weekend off once, I took a group of students down to our bungalow at Betty's Bay.

On a particular Saturday afternoon there was a rugby match between England and France which the youngsters wanted to see, so I took them down to the local pub which has a large screen for just such occasions. Two of the students had no desire whatsoever to watch the match and, because the Springboks weren't involved, neither did I, so I took them for a drive around the area. At Pringle Bay I could see two whales close to the rocks and, without telling the students, drove as close to the area as possible and then told them to follow me down to the rocks. I was waiting for them to see the whales, but even when we were right there, they didn't see them. I pointed at the large grey mass of surfacing whale in the water and said, "Can you see the whales?" They both said no, they couldn't. I couldn't believe that right there, just 25 metres away, they couldn't see the two floating whales!

It was only when I looked at them that I realised that they were looking far out to sea and when I said "look down" that I had the privilege of seeing one of the most uninhibited joyous reactions I've ever seen. The male student literally jumped into the air and didn't stop hopping around and yelling, the girl started shaking with excitement and incomprehensible words bubbled from her. I was rather concerned at first till I realised that it was unabashed joy that I was witnessing and we spent the next hour watching these two behemoths mating right in front of us.

It is occasions like this that really give me enormous pleasure, watching people see something for the first time in their lives and knowing that it will live with them forever. It goes without saying that when we collected the others from the pub, some of them regretted not having joined us. This regret was coupled with the fact that France had beaten England in the rugby.

CHAPTER 28

LEOPARD

Leopards are very rare in the western Cape as they've been persecuted to such an extent over the years that they have literally become mountain leopards. Some people even call them the Cape Mountain Leopard. Their territories are vast and in some areas the male territories have been measured at close to 500 square kilometres. They move high on the ridges of the mountains, away from human activity and so are very rarely seen, just the occasional pug mark in a sandy river line or some scat or claw marks on a tree. They are incredibly secretive and, till recently, very little was known of their movements and habits. This is changing rapidly with new technology. There is a specific leopard that has been tracked by means of a satellite tracking system which can be downloaded when the animal is within range of the receiver. The resulting information has been nothing short of astounding.

The leopard found in the Cape appears to be smaller than its relatives to the north but makes up for its size difference by being incredibly mobile. Our specific leopard that has been monitored for a few years now has a range that is so big it actually covers an area of over half a million hectares. Having said that, he uses a route that is approximately 340 kilometres long, roaming the ridges and valley floors and rarely enters the flat areas.

It would be very interesting to see the research into these animals' dietary habits, as up on those mountains prey is small and doesn't occur in large numbers. It will also be interesting in the future when more than one animal and also animals of different sexes, can be monitored as it will give far more information.

On a shelf in my office is the skull of an adult male leopard. It is a grim reminder of the fate of many leopards who stray into farming areas, especially where stock farming takes place. Just when we started the reserve all those years ago, I received a call from a friend at Cape Nature Conservation asking me if I could assist in the trapping of a problem leopard in the area. I immediately volunteered and got the team and our equipment together and headed off to the farm.

On arrival we were met by a very shaken farmer and it turned out that he had gone to see his problem animal traps and instead of trapping a jackal or a caracal, he had inadvertently caught a large male leopard who, by this time, was injured and angry. Without hesitation the leopard, cornered and unable to flee, turned his attention to the farmer and his dogs, so the farmer had little choice but to shoot the animal. So even though we were too late to help, the trip wasn't a complete failure as I had time to discuss the situation with him. He gave us his word that from that moment on he wouldn't use gin traps again and we helped him source some fall traps. Over the years we received calls from him telling us what he had trapped and we helped him find release sites for various species.

We took the leopard carcass through to the officials and I asked them if it was possible, when they were finished with their investigation, if I could acquire the skull for teaching purposes, and over the years many students have learned the anatomy of a leopard's head from that same skull.

Rarely, one will be trapped in a cage trap set out by a farmer who is losing domestic livestock to predators. On three occasions we received live leopards from Nature Conservation officials who had caught a problem animal and needed a safe place to release it. We were totally unprepared for the first one that arrived and didn't have any equipment to do any research on it, so we released it as quickly as possible into the reserve in an area that had all of the habitat requirements it needed to survive. The trap that it had been caught in was designed for caracal and added to that it was incredibly old and rusty. I suspect that it would have easily escaped had there not been a cover placed over the trap making it dark inside. I'll never forget the way that this release was handled and all I can say is that it was a miracle that nobody was torn to shreds.

The Nature Conservation official had stood on top of the cage with a shotgun at the ready and raised the falling trapdoor. Fortunately for everybody the leopard took one look at the mountains and bolted for them. Leopards are so fast it's difficult to describe it, and when in confined areas, the speed at which they can react is truly frightening; one wrong move can mean your last.

So when the second leopard arrived we were prepared – we had acquired a spare tracking collar and the right drugs to do a controlled release. The female leopard arrived on a very cold, wet evening and, after inspecting the cage, I decided to anaesthetise her immediately as she had nasty abrasions to her nose from repeatedly bumping into the cage when she went on the offensive. We could also keep her warm during the night while she regained consciousness and release her in the early morning when the effects of the drug had worn off. The dart gun that we use can't be used at close range so we had to use a hand-held syringe, so with Paul at one end distracting her and me at the other end with a syringe waiting for her to press her rump up against the cage, we waited for the opportunity. It took a while but we eventually got a chance to give her a jab. The speed at which she reacted really got my attention and reminded me just how careful one has to be when dealing with a captive leopard.

Once she went under we gave her a thorough check over – she was in very good condition and just the abrasions and swelling of her nose was of concern. This we treated with various medicines and after giving her a general purpose antibiotic, we put on the tracking collar, made sure that she was safely in her cage, and in a warm environment with blankets and covers. We left her to come around in her own time and the next morning when we inspected her again, she was just as feisty as when she'd arrived and the piece of springbok meat that had been left in her cage was gone, with her water bowl empty. In my opinion it's always a very good sign when an animal takes food and water and it also means that if the leopard is being released into a new environment, they have enough sustenance in them until they find water and can hunt for themselves.

We collected all of the equipment needed and went to a quiet spot with prey and water nearby, but more importantly, in a safe environment that she was accustomed to, right at the base of a mountain. We set the cage facing the mountain, in such a way that Paul could use a rope to open the cage from the safety of his vehicle about 30 metres away. When everything was set up and ready, I went around to all of the vehicles full of students and staff and asked them to close their windows and to make sure that they stayed that way till the animal was well away from the vehicles. Having released a few leopards before and knowing what they can do, I did the round again and made sure that everybody had closed the doors and windows, then climbed into my vehicle and gave Paul the go ahead. He pulled on the rope and the door to the cage fell away. Nothing happened for about 15 seconds, then the leopard's head peered out, first looking left, then straight ahead and then right to where the vehicles were.

Then all hell broke loose. It went crazy, followed the rope to Paul's vehicle and flew at the windscreen, then slipped down the driver's side door and bit the outside rear-view mirror. Then it dropped onto the ground, bit a hole in the tyre, which immediately deflated, then crawled underneath the vehicle, came up at the passenger door, onto the roof and slid down the bonnet before running away up into the hills. This all took place in a matter of seconds. I can tell you that there wasn't a person at that release who will ever doubt the capabilities of an angry leopard. Paul's father had been seated in the passenger seat of the vehicle and had unwisely lowered the window a little. When the dust had cleared and we were all standing around reliving the event with nervous amazement, he told us that he wound up the window so fast that the handle had come off in his hand!

The leopard, under normal conditions, is not a danger to man at all and will always avoid human activity, but when humans trap them they become incredibly defensive and will attack when subjected to this kind of stress.

CHAPTER 29

MORE TRAVELS

I was on a visit to the UK and France for both personal and business issues. Our daughter Janet lives in London and we miss her terribly so at every possible opportunity we get, we try to visit her. I was giving a talk to a group of students who were going to visit the reserve, and I'd been invited to give a lecture at London University which, I must say, was a great honour. A few days later I was to give a talk to a group of VIPs at South Africa House. I am one of those people who are very aware that technology in one part of the world does not necessarily have a good working relationship with technology from another, so I asked the folk who had invited me if I could check that my equipment was compatible with their system before the event.

When we arrived at South Africa House that morning we were met by a wonderful gentleman who helped me check the equipment, which was just as well because there were a few issues that had to be sorted out before it all started working in harmony. He could see that Simmy and I were intrigued by the building and he kindly gave us the grand tour. The most amazing history is on the walls of that building and he took us to all of the places that are not normally shown to the general public as it was a Saturday morning and most of the staff were off for the weekend. There is art work by some of the most famous artists that South Africa has ever produced, from works that date back a couple of centuries to some of the most modern works of today's artists. We even went out onto the top balcony and had a bird's eye view of Trafalgar Square where they were having a ceremony unveiling one of the statues.

That evening I was greatly privileged to have Dr Ian Player introduce me at the South Africa House talk – to me a truly great honour. This legendary conservationist had just arrived in London and heard about the talk that I was giving and asked if he could attend. I felt really privileged; it's something I will always cherish. It was an evening of a couple of firsts for me – my mother was in London at the time and this was the first time she would hear me give a lecture.

Ian Player is a great man, a man who has lived an incredible life dedicated to conservation. Wherever he goes he quietly gets the conservation message across, changing attitudes and creating awareness of the plight of so many of our plants and animals. His honouring me in such a way made me more determined than ever to raise conservation awareness wherever I go. The talk was a great success and quite a lot of money was raised for the Wilderness Foundation.

It was also a time to see old friends and folk that I had worked with over the years. I'd arranged to revisit Bob

Lawrence at the West Midland Safari Park near Kidderminster, outside Birmingham. Bob had started a white lion breeding project at the Safari Park and I was keen to harvest DNA samples from his lions. I'm still of the opinion that most of the white lions in captivity today are related and all of the interbreeding is going to create problems in the future.

A couple of years earlier I had been visiting Alison and David, my cousins in Kidderminster, and David who is a Scotsman with a wicked sense of humour, told me that there were white lions at Bewdley, just outside Kidderminster. It was April the first so I just laughed at him and the more he tried to convince me that it was the case, the more I kept saying that his April fool's joke wasn't working. Added to that was the fact that literally every second pub in the UK is called the White Lion, so I was suspicious that the white lion he was talking about was some local watering hole. Only when Cousin Alison produced a brochure from the park did I realise that he wasn't trying to pull my leg. With that we phoned and made an appointment to meet with Bob and after our first meeting we've kept in contact ever since.

I have a sample of harvested DNA from all of the lions that I have worked with and one day – when I can get the funds together – I'm going to have them analysed to check if my suspicions about inbreeding are correct. It was good to see Bob and I can say without any hesitation that there is no other captive facility that I have ever been to that looks after the welfare of its lions better. Bob and his team are a shining example for the rest of the zoos and safari parks around the world. I'm not in general a great supporter of captive facilities; however being a realist, one has to understand that most people in the world today can't afford to travel. So I'm grateful for places like Howlett's, Port Lympne and West Midlands Safari Park that create awareness of the plight of our animals around the world, with their excellent education programmes. Some of them even have re-introduction programmes that are working in many parts of the world.

CHAPTER 30

Hard Decisions

Whenever I fly into Cape Town and see Table Mountain from the air, I just know like most Capetonians, that I am home. It is a remarkable geological formation rising out of the sea and although it looks flat, speak to anyone who has walked up there – it's anything but! In the San language it is called Hoerikwaggo, which literally means 'Mountain in the Sea.'

On my arrival back at the reserve after my overseas trip, some significant decisions had to be made regarding all the lions and how the programme was to go forward. With the one white sub-adult male still causing problems with the game drives, and any vehicle for that matter, a decision had to be made before any nasty situations arose.

His actions were not at all aggressive; in fact it seemed to me that it was mostly a combination of curiosity and youthful spirits. However, any animal that loses fear of vehicles and humans is more than likely to become a danger to the safety of the guests. It's rather daunting to have a lion approach right up to a vehicle but when it actually wants to get inside with you it's a totally different story.

No matter that his intentions might be to give you a big lick... After trying many methods to keep him from approaching the vehicles had failed, it left us with no option but to remove him from the wild programme. It was also nearly time to separate the cubs from their parents as they were approaching two years of age, and fighting between Jabulani and his sons would soon erupt. Under natural conditions the young males would be moving away from the pride anyway.

As we were getting our future plans in order, the reserve got a message that Linda Tucker was going to collect her three lions that we'd been housing. This would ease the strain on boma feeding as that is one of the jobs that is really tough for even the most dedicated conservationist. To go out up to three times a week and have to shoot animals that you really care for is not an easy task to expect from anybody. I'd tried to share the task between Ryno, Paul and myself but due to other pressing requirements, Ryno was the one who carried this burden the most. Fortunately, he's an excellent shot and did this work with skill and concentration.

Linda arrived and they flew the lions out to their new home. Not long after their departure I heard from a friend that the adult female had died which was very sad news because she was a very fine mother and had raised her cubs very well indeed. She had the most incredible eyes – they were halfway between green and brown with dark brown speckles. She was also a very large lioness with a stable temperament which made her a pleasure to look after for all the time that we had her. With her very good mothering instincts she was an ideal candidate for the continuation of the white lion gene. After Linda had collected her lions, the pressure of culling for predator food became a lot lighter and it freed the team up to do other important work and to make certain decisions.

The decision was taken to remove Jabulani and Queen to a separate area and to continue with their offspring, the three males. This was done and the team once again had the chance to give them a full physical examination. They were both in very good shape, and the timing of their removal had been spot on because the aggression between Jabulani and the youngsters had just started to leave marks and scars on their faces. The problem with the one youngster, however, didn't stop and he really was becoming a very big concern to all as he continually approached any vehicle, both game drive vehicles and the Wildlife Departments bakkies. (The word bakkie is used by everybody in South Africa for a small pickup truck which can either be single or double cab with an open container at the back for loading.)

The time had come to start the integration process. The pride of tawny lions on the reserve was doing well so the decision to take two of the females and bond them with the three white sub-adult males was made. The methodology would be to put the three males into one boma and the two tawny females in the one next door and see what would happen. This was done and the usual posturing took place but no real aggressive behaviour was seen. After a week or so the lions were seen lying down next to the fence close to the females and no aggression was observed at all. The decision was not to rush the bonding process and they were left in their separate bomas for more than 30 days so that they were very familiar with one another in many ways. They could see one another all the time because they would lie down right next to each other beside the fence; their smell would also become familiar and they could vocalise.

The bonding process worked very well because when they were introduced there was very little aggression. The males were still young and this was novel to them and they were inquisitive about the new members. The females, however, gave a good show but it was mostly posturing and they soon calmed down and accepted each other, with normal dominant displays by the males when they fed.

They were released into an area of about 5,000 hectares that had been stocked to capacity with prey items. There was never any worry about the females coping; they were both incredibly successful hunters and had thrived on the reserve. The question was: what would the males do? It didn't take long to find out – they became typical male lions and followed the females closely. When they made a kill the lionesses would let the males take their share first without any aggression, so the males would often be seen together but never very far away from the females. The bonding process was working and they started to display typical pride behaviour and even though the males were still sub-adult, sporadic mating did take place.

Queen and Jabulani were put into a separate boma far away from the new pride and went about being the relaxed lions that they were, so it was actually quite a pleasant surprise to find (a while later) three new cubs in their enclosure. It wasn't entirely unexpected but it was still a wonderful situation because it now created the possibility of having two separate prides on the reserve. After a few weeks it became possible to sex the cubs and it was another interesting surprise to find that they were two females and one male. These were the first females born on the reserve so it opened up the options for the way forward.

The new pride was observed on a daily basis and as the three males started maturing so the visual impact of having three male lions as part of a pride grew. It was truly an impressive sight to see three male white lions together with the two tawny females, truly the first free-roaming white lions for many years.

CHAPTER 31

RHINO

The reserve kept expanding, which meant more staff and more responsibilities to attend to. Guest numbers were up and the Wildlife Department was fortunate to get some extra people on board to do the necessary monitoring. With so many new species arriving, the monitoring of the game became a very big concern to all. We had a black rhino in the south that was such an elusive creature that we called him Bin Laden. One of my favourite things to do is track black rhino on foot. Often on Sundays, I would sneak off and track the rhino with the telemetry set. When he arrived his horn was cut back and a transmitter was put in the stump of his horn. The telemetry set looks a bit like an old television aerial attached by a cord to a box that you can set to different frequencies. The idea is that you point the aerial and if it picks up a signal it starts beeping, and as you get closer the beeping gets louder. The reserve is made up of hills and dales and the telemetry doesn't react well to mountains which means that you can get all sorts of bounced signals which we got used to and compensated for. In fact, the team became highly proficient in using the system. So one Sunday I was out looking for the rhino and got a really good signal telling me that I was within about 100 metres of him.

With black rhino having the most incredible hearing, it was common practice to switch off the telemetry when getting nearer as they could pick up the beeping sound. It was then up to the tracker to get sight of the animal before he became aware of you. I slowly and quietly started walking in the direction of the signal and after about 100 metres still couldn't see him. I walked around some more and still couldn't spot him so I switched on the telemetry. The signal was really loud and I hadn't even raised the antenna. I turned the volume down as low as possible and scanned the area; the signal was coming from very close by and I was out in the open on a ridge with very few places that a rhino could hide. The signal was coming from a ridge in front of me so I decided to slowly walk forward. As I went over the ridge I saw him, fast asleep in the afternoon sun less than ten metres away.

I reversed very slowly and carefully back to a Ghwarrie tree (*Euclea undulata*) and sat down to observe him. He was totally unaware of my presence and what was really special for me was that I could hear him snoring. I must have got into a very relaxed state because when the time came for me to leave, I accidentally dislodged a stone. It's at times like these that you find out just exactly how fast a rhino is. In the blink of an eye he transformed from a sleepy, snoring rhino to a very alert and wary chap. I stood absolutely dead still knowing that his eyesight wasn't that keen, and the fact that

I had a tree behind me, breaking my silhouette, would make it difficult for him to see me.

Black rhino, when agitated, do a side to side step, almost like a little rocking dance which can look quite comical, unless it happens to be less than 20 metres away from you and the only tree near you is rather flimsy and only just over two metres tall. They also snort and grunt which, combined with their lightning-fast reactions and stomping actions, make them rather awe-inspiring to say the least. They seem to be able to pirouette fast in any direction and, for their size, they are incredibly light footed. I stood glued to the spot, absolutely awe inspired while he gave this demonstration of his capabilities, and no sooner had it started than he relaxed and lay down again and went back to sleep. This time I was very aware of any rocks in my way and sneaked away leaving him to his siesta in the warm winter sun.

I used to love the time of the full moon. I would often walk on the roads on cloudless, full moon nights. The walk up to the runway was about five kilometres and some nights it felt like walking under a spotlight. The quiet of the night, free of light pollution and human sound is a wonderful experience which I would share with some of the students. It was a time to just be a part of a magical environment. We would often hear elephants going about their lives and, more often than not, a lion roaring somewhere on the reserve. When we went through the riverine area the kudu would give a loud bark, warning of our approach. There was a total eclipse of the moon at about midnight one night and we all bundled into the vehicles and drove up to the high point next to the runway. It was incredible to watch this phenomenon in the Karoo where the air is so crisp and clear.

This has all changed, however and I now dread the full moon. It brings a sense of trepidation and anxiety to me and it makes me angry, frustrated and helpless. Why? Because I know that on the news the day after full moon I will hear or read that the poachers have struck again. Rhino poaching has taken on unbelievable proportions in our country. During the 1960s estimates of the black rhino population in the rest of Africa, excluding South Africa, varied from 50,000 to 100,000 and the northern white rhino 2,000 to 4,000. Over the next 50 years the world completely lost the northern white rhino and there are less than 2,000 black rhinos in the rest of Africa. In South Africa in 2011 we lost 443 rhinos to poaching, a total which has already been passed two-thirds of the way into 2012. It is now a sad fact of life that if this is allowed to continue we will lose all of our rhinos. The system that the world and South Africa has in place isn't working, which is proved by the fact that we're still losing animals daily. Rhino owners are scared because the poachers are ruthless killers who aren't averse to killing the land owners, the national park rangers or anybody else, to get at the rhinos.

Many of the private owners are trying to sell their rhinos but of course nobody wants to buy an animal that can bring about murder and mayhem to their property. At present it is illegal to trade rhino horn, which has pushed the price through the roof, making it very attractive to poachers and dealers alike. When there are such vast amounts of money being paid, people are prepared to go to extreme lengths to be a part of it, even if it can be life-threatening.

Historical uses for rhino horn date back thousands of years. In Greek mythology, they were said to possess the ability to purify water. The ancient Persians thought that vessels carved from the horn could be used to detect poisoned liquids, a belief that persisted into the 18th and 19th centuries among the royal courts of Europe. In the Middle Eastern country of Yemen, the horn continues to be coveted by Muslim men, even though it became illegal there to import it in 1982. The material, whose lustre increases with age, is used for the handles of curved daggers called 'jambiya,' which are presented to Yemeni boys at age 12. Jambiya are considered a sign of manhood and devotion to the Muslim religion, and are used for personal defence. Yemeni men place great value on the dagger handles, which are often ornately carved and studded with jewels.

Far more widespread is their use in the traditional medicine systems of many Asian countries -from Malaysia and South Korea to India and China- to cure a variety of ailments. It is also, mistakenly, popularly thought to have aphrodisiac effects. In traditional Chinese medicine the horn is shaved or ground into a powder and dissolved in boiling water, then used to treat fever, rheumatism, gout and other disorders. Historically it was also used to cure snakebites, hallucinations, typhoid, headaches, carbuncles, vomiting and food poisoning. Rhino horns are not, as once believed, made simply from a clump of compressed or modified hair. They are, in fact, similar in structure to horses' hooves, turtle beaks and cockatoo bills. Rhino horns are composed largely of calcium, melanin and the protein keratin which just happens to be the chief component of human hair and fingernails.

All five of the world's species of rhino have been brought to the edge of extinction because of human appetite for their distinctive horns. Fortunately, science is able to dispel the fiction surrounding rhino horn. Science is now stepping in to dispel some of the mystery and myths surrounding its use. Unlike the horns of most animals, which have a bony core covered by a relatively thin layer of keratin, rhino horns are keratin all the way through, although the precise chemical composition of the keratin will vary depending on the rhino's diet and geographic location.

It is believed there may be some truth behind the rhino horn's ability to detect poisons which is linked to the composition of the horn. Many poisons are strongly alkaline and may react chemically with the keratin. With this information we can now take 'fingerprints' of the horns and determine where the animal populations came from, which has helped law enforcement officials target some of the areas where poaching is taking place.

Overall there isn't much evidence to support the plethora of claims about the healing properties of the horns. In 1990, researchers at the Chinese University in Hong Kong found that large doses of rhino horn extract could slightly lower fever in rats (as could extracts from Saiga antelope and water buffalo horn), but the concentration of horn given by a traditional Chinese medicine specialist is many times lower than that used in those experiments. I really wish with all my being that the people who use rhino horn, for whatever medicinal purpose, could understand that they would do just as well chewing their fingernails.

The international trade in rhino horn was banned in 1976 by signatories to the Convention on the International Trade in Endangered Species (CITES). In 1993 the Chinese government also banned the use of rhino horn, or any other parts of the body of the endangered species, in traditional Chinese medicine. However, as we've seen, this system isn't working and the animals will become extinct fairly soon unless some 'out of the box' thinking is done to stop the carnage. I've been in contact with people who are advocating some very workable solutions that would not only benefit the rhino but also benefit poorer communities who would be able to legally harvest the horn.

What is not often explained is that you don't have to kill a rhino to take its horn. The dehorning of rhino is now a standard anti-poaching method widely used by many private and national parks. The horn grows back at a rate of up to 12 cm each year with the horns of females growing slightly faster than those of males. Horns can therefore grow back to their original length and shape in time, but only if the growth plate of the horn isn't damaged. It has been suggested that flooding the market with horns harvested from dehorned rhinos would reduce incentives for poachers. Knowing that rhino horn can be obtained without killing the animal, would represent a renewable resource. Plus there is also the fact that considerable quantities of confiscated and legally harvested horns are building up in warehouses, so if they were sold legally it would create a reduction in the global price of rhino horn and most importantly, reduce the incentive for killing them.

There is a case study from Peru on the vicuna, which is a member of the camel family and distant cousin of the llama. Simply put, vicuna fleece is the rarest, most expensive natural fibre in the world. Its insulating properties make it warmer than wool. Unfortunately, due to the high value of the fleece, vicunas were nearly poached to extinction, and by the 1960s it was estimated that there were fewer than 6,000 left. They were being killed for their wool which could have been sheared every two years without killing the animal. Does this sound familiar??!!

The government of Peru and private enterprise have now turned the situation around, where the local population now get the opportunity to round them up and shear the wool every second year, bringing in a sustainable revenue without killing the animals. Since that time there has been very little poaching because the people who live out in the areas where the vicunas are found, protect them. They now have a viable business and the only vicuna wool that can be legally sold comes from these Andean farmers. There are now more than 370,000 vicunas and they have been removed from the Endangered Species List to a least threat list.

With this success story as a basis, I'm sure that the thinkers of today could adapt it to the rhino situation, because the present laws and regulations are not working. Over the years, the Zimbabwean authorities embarked on a very extensive dehorning exercise and they emphatically stated that they couldn't detect any negative effect on the social behaviour of the dehorned rhinos. I'm sure there is enough research already to cover this aspect, so my appeal to all concerned is as follows: Stop the politicking and the infighting and sit down and create a solution that will work, so that our rhino population will not become extinct.

The Chinese and Asian people are not going to change. A wealthy, influential politician even told the people that his life was saved by drinking water with powdered rhino horn stirred in it. We have to use a different approach. Please believe me, the last thing I want to see when I walk around the bush is a rhino without horns; however, I would rather see a live rhino without a horn than a dead one that has been butchered. I have just read an article in a magazine about an Irish gang that is stealing rhino horns from museums and collections around Europe. Many of the museums have now removed the horns from their specimens and replaced them with wooden, glass fibre and plastic ones. Security at zoos and wildlife parks has been upgraded to protect their rhinos. Germany, Spain, Belgium, Portugal, Switzerland, Italy, France, Sweden, Austria and the UK have had a total of 58 thefts, resulting in the loss of 72 rhino horns. Nothing short of a profound shift in eastern culture will make our rhinos safe.

CHAPTER 32

CHANGES

The three young white males and the two tawny females did exceedingly well in their environment; however, the one male was still a problem around vehicles. The snag with this was that people didn't just want to hear about the lions, they wanted to see them. A decision had to be made in this regard and it was also important that this kind of behaviour shouldn't happen to the next litter that had just arrived. There was no doubt that even though every effort had been made to not let the three founder cubs become habituated to vehicles and people, the one had done just that. He was too inquisitive and, at that young age, didn't have the natural fear of vehicles that normal lions have.

Hindsight is an exact art and the conclusion was that we'd waited too long before introducing them to the tawny females. With the new arrivals we would have a better chance of them not becoming habituated if they were integrated and bonded with wild lions at a younger age. The decision was to integrate them at an earlier age so that the influence of

Jabulani and Queen would be lost sooner. The best time for this would be after they'd been weaned at about six months.

This was the time when the difficult decision had to be made as to what to do with the problem white male in the pride. After great deliberation and many backward and forward discussions, it was decided he should be sent to Shamwari in the eastern Cape. They had all of the correct facilities and it would relieve the team of the stress that he was creating around the vehicles and people.

A new method of bonding was also going to be attempted on the next habituation attempt. The decision was to take the two tawny females away from the three males at the same time as the removal of the problem male. We would leave the remaining two males in their area to carry on with their lives, while the two females would be bonded with the three new cubs. This was going to be carried out on the main reserve so that the two males could join them at a later stage. This bonding process was slightly different in that the vegetation in the bomas was such that there were trees for shade but the lower vegetation was scrub so that literally everything inside was visible, except the feeding platform at the back.

Within days the three white cubs had settled down in their boma and right next door, the two tawny females were showing signs of interest in the three youngsters. It wasn't very long before they were all lying next to the fence communicating with each other. They were very relaxed in each other's company and the decision was made to introduce them and bond them as soon as possible. This is where the new method would be used. For two days before D-day no food was given. They were all sedated and while unconscious, the two adult females were given a long-acting sedative to keep them calm. A large carcass was put into the boma so that when they woke up the idea was that they would be slightly hungry and this would concentrate their efforts on eating and reduce the chances of tension and fighting breaking out. It worked very well and within days the cubs and the tawny females were interacting in the form of grooming and playing together. The youngsters were even accepted at the carcasses at feeding time. It didn't take long to realise that this integration and bonding had been a success and the way forward would be the release on the main reserve.

Next, the tawny pride on the reserve would be captured and put into the boma and the two tawny females with the three integrated cubs released. It would be looking for trouble to release them onto the reserve with an existing pride. At the time, there was a female with four very young cubs so the decision was to leave her out on the reserve and capture the rest of the lions. The release went very well and the two females hunted and let the cubs feed so the project was starting to look as if it was on track.

Just when things really started to look as if they were settling down, the tawny female with the four cubs came across the new pride and a skirmish ensued which left the young white male dead. This is the way of lions, totally unpredictable, and in human terms absolutely brutal. It was imperative to get the rest of the integrated pride to safety as quickly as possible so they were captured and placed back in their boma. The project had to be re-evaluated and time was needed to get everything on track again, so the original tawny pride was released back onto the reserve. This was not a good time – after the successes that had been achieved, to lose the young male was a very big setback. The way forward would have to be planned in such a way that this wouldn't ever happen again.

After lots of very intense discussions and debate it was decided to relook at the two sub-adult white males again. They had been out on their own and had hunted everything from giraffe to buffalo and many different species of smaller game. The most important factor was that since their brother had been removed they were not the slightest bit interested in vehicles or people. So a decision was made to put these two male white lions into the boma next door to the integrated tawny females and the two juvenile white females, and observe what would happen.

After the introduction there was the usual posturing and posing but no aggression whatsoever, and within minutes they were lying down next to the fence within touching distance of each other. The two young white females were completely at ease with the two males and one of the tawny females starting showing signs of oestrus.

The decision was to give them a month and observe them on a daily basis and if all went well the introduction would be done. The other decision was that all other lions on the reserve would be translocated to Shamwari so that there would be no chance of any interaction between rival prides, and that when the new pride was released, they could just concentrate on becoming a cohesive unit. It became clear that there was a very good chance that this integration would work; all of the reactions between the two bomas were calm and relaxed and a day was decided upon to do the introduction.

The same method that had worked so well before was followed and when all of the lions had been darted, the two males were put into the larger boma with the two tawny females and the two white cubs. The adult animals all received a long-lasting tranquilliser and a kudu carcass awaited them when they woke up. It was a great relief for the whole team when they all came around and no aggression was seen and they all tucked into the kudu. Things certainly started looking up again as the team had worked so hard and had been beset by setbacks; we welcomed the fact that things finally looked as if they were going to go well again. The date for the release onto the main reserve was set and it was vital to the success of the project that the monitoring system was in place and that all parties were aware of what was going to happen.

When all of the other lions had been removed from the reserve, the new integrated pride consisting of the two sub-adult white males, two adult tawny females and two white female cubs, was released. Sanbona now had one pride of lions; however, nothing about lions is predictable and the wildlife team was on high alert for any possible incident.

Weeks went by and the monitoring team kept a very tight schedule observing the lions and rejoicing that they weren't having to deal with dramas and problems. The pride settled down to lion everyday life which was hunt, eat, drink and walk, and even a bit of mating when one of the tawny females came into oestrus. Guests were having some amazing viewing experiences and the team, for the first time, started to relax about the lions. This was just as well because at this time the reserve was being sold to new owners and with it came some teething troubles and some new thinking.

While all of this was going on, the functioning of the reserve continued and white rhinos, giraffe, extra buffalo and many antelope species numbers were introduced. New lodges were built and the Gondwana student centre was completely revamped into a lodge – and sadly the student programme was stopped. I must say that the idea of not having a student programme on the reserve was very sad indeed as I'd really enjoyed having them, even the ones who were completely crazy. The programme brought people from all over the world and from all walks of life and gave them the opportunity to learn about Africa, its wildlife, floral kingdoms and people. I learned a lot from the students – there were times when they would ask questions that really made me think very deeply about what we were doing and how we were doing it. It was sometimes a pain to be asked the same question over and over again; however, more often than not we would get questions that demanded minute details on a subject that we loved; and to see the reaction when a young person came to understanding was a privilege for me. Fortunately, I have kept corresponding with many of the students and keep up to date with what is happening in their lives. Many are now qualified in their chosen subject and some are even running their own successful businesses.

CHAPTER 33

African Wild Dogs

Somebody at head office got a bee in their bonnet about introducing a family of wild dogs onto the reserve. Wild dogs are southern Africa's most endangered large carnivore and one of the most endangered carnivore species in the world. Wild dogs hunt in packs and small groups. Like most members of the dog family, it is a cursorial hunter; this means that they pursue their prey in a long, open chase. They are possibly the most successful carnivores because nearly 80 per cent of all wild dog hunts end in a kill. To put that in proper perspective, lions, who are incredible predators, only have a success rate of about 30 per cent. In central Africa a study showed that 90 per cent of wild dog hunts ended in kills. The members of a pack vocalize to help coordinate their movements. The voice is characterized by an unusual chirping or squeaking sound, similar to a bird.

It must be understood from the outset that due to the nature of the environment of Sanbona, the population densities of the herbivores must be low. With low rainfall and being an arid zone, the vegetation is susceptible to being overgrazed. To put it as simply as possible, the quantity of prey dictates the numbers of predators – which is called the predator/prey relationship. After many discussions the ruling went against the reserve and the dogs were going to be sent to us, despite all of the objections.

The boma was prepared and the day arrived. Five dogs were put inside for their acclimatisation phase. They had previously been in a boma and had, like the cheetahs, been feed-trained with a whistle. Well, you can imagine the shock when the team arrived the next morning to find an empty boma with an escape tunnel as a new feature. From that point on it was like a killing spree – on the first day they killed two springbok and a young kudu. After five weeks of exhausting monitoring and watching the prey species dwindle to the point that was just impossible to believe, a decision had to be taken before there were no antelope left on the reserve for the established lion, cheetah and hyaena populations. The dogs were darted and sent back home to Shamwari – they were just too successful even with 55,000 hectares to run around in. I estimate that a reserve would have to be close on 250,000 hectares to sustain a wild dog population.

I read an interesting comparison recently that a wild dog has the constitution and metabolism of seven sheepdogs. If you have ever owned or had dealings with sheepdogs you'll know that they just don't stop, so you can imagine what seven all bundled into one would be like!

CHAPTER 34

BUFFALO

Buffalo can be carriers of the dreaded foot and mouth disease which is sometimes known as hoof and mouth disease (*Aphtae epizooticae*). This is a highly infectious and sometimes fatal, viral disease that affects cloven-hoofed animals, including domestic and wild bovids. The virus causes a high fever for two or three days, followed by blisters inside the mouth and on the feet that can rupture and cause lameness. Foot and mouth disease is a severe problem for animal farmers because being highly infectious, it can be spread by infected animals through the air, through contact with contaminated equipment, vehicles, clothing or feed, and by domestic and wild predators. In South Africa buffalo are a great financial investment for reserves that fall out of the foot and mouth area. Of the eighteen southern African wildlife species known to have antibodies to foot and mouth disease, only the African buffalo has been shown to be an efficient maintenance host of the virus and it is considered to be the primary source of infection for domestic and wild ungulates. This is due to its ability to both maintain and transmit the disease. In simple terms our buffalo can live normal lives and not show any adverse effects from the virus. However, they transmit the disease.

Current legislation in South Africa specifically requires all buffalo to be free from the following diseases before a group or individual animal is moved from its current location: Foot and mouth disease, tuberculosis, brucellosis and corridor disease. Foot and mouth disease can cause great economic losses and hence, not only is it managed locally, but internationally by the World Organisation for Animal Health, the OIE. Strict monitoring and protocols are followed in a foot and mouth disease zone and this ensures that the disease is not transferred from one area to another.

There are normally three zones in place – the Red zone which is the infected area, the Yellow, which is the fascination zone and the Blue, which is called the monitoring zone. People wanting to move buffalo will first capture the animals and place them in a quarantine facility. Then blood samples will be taken and sent off and tested for the four diseases, during which time the animals remain in isolation till the results are available. When animals are declared clean, the details of the animals will be taken and recorded by the State Veterinarian Service Department and micro-chips will be installed in the buffalo to ensure animals can be traced and identified.

Only once this is done can one apply for a permit to transport the animals. So, what has happened is that disease-free or 'clean' buffalo, have become a very sought-after species and come with a very high price tag. At an auction one buffalo sold for $2,500,000! Yes, that's correct – two and half million US dollars. The demand for buffalo outside of the foot and mouth areas is high which has created a thriving trade in the wildlife industry with disease-free buffalo.

With the introduction of our first herd into the reserve, the buffalo were released straight from the truck into the open. They had already spent a couple of months in bomas where they had undergone all of the required testing; we were informed that they wouldn't need any more time to habituate to our environment and that they must be released onto the greater reserve straight from the transport vehicle. Well, this was a great mistake if ever there was one! They were released in the area around the dam and within a day nobody could find any trace of them anywhere. We sent teams all over the reserve to search and all came back with not a single sign of them, so to all intents and purposes they had simply vanished. As a last resort we decided to search in the mountains, and you can imagine our absolute amazement at finding these normally plain-dwelling creatures high on top of a mountain range, munching on fynbos and reeds.

The term 'fynbos' has been used since the time of the Dutch settlement at the Cape. The predominant vegetation on the Cape Peninsula had timber too fine and slender for harvesting and was thus given the name 'fijnbosch.' The name may also refer to the dominance of small or fine-leafed plants. Fynbos makes up 80 per cent of the Cape floral kingdom, which comprises an area of 90,000 square kilometres but hosts a staggering 8,600 species. To put this in some form of perspective, the British Isles, three and a half times larger, have only 1,500 species and less than 20 of those are endemic. Table Mountain alone has almost 1,500 species in 57 square kilometres.

Dry mountain fynbos occurs on the ridges and higher areas of the reserve, where there are four distinct growth forms: the proteoids, which are tall protea shrubs with large leaves; the ericoids which are heath-like shrubs; reed-like plants called restios and bulbous herbs known as geophytes. Fynbos is not high value grazing at all so when the buffalo were found grazing contentedly on the reeds and ericas, I was pleasantly surprised. Knowing that very few of the animals utilised the fynbos on the ridges it was good to know that it was going to supply sustenance to the buffalo. However, this did come with one drawback – as there were no roads up to the ridges, it meant that only when they came down for water would the guests have an opportunity of seeing them.

Sadly, we suffered some losses from the first introduction but the herd slowly started to grow over the years, until the lions developed a taste for buffalo steak tartare and started decimating the herd. A decision was made to capture the survivors and put them in a breeding area to build up their numbers till the herd was big enough to sustain predation.

Another aspect that caused quite a lot of headaches was the introduction of giraffe. Historical records show that they had lived in the Cape at one time but no records by any of the pioneers and early travellers gave any reference to their living in the Klein Karoo. Because of this they are described as an extra-limital species for the area. I'm not exactly sure why this decision was taken as there is certainly no reason why they couldn't have occurred here. Be that as it may, the introduction of giraffe to the reserve has been a resounding success. They have bred on the reserve and the offspring have also successfully bred, so at this point they are doing very well indeed.

In the Kgalagadi I've seen and even photographed gemsbok (oryx) chewing on bones, which is called Pica, or more specifically, osteophagia. I had before that never witnessed a giraffe chewing on bones so you can imagine my surprise when I was right out on the eastern boundary filming the giraffes to confirm their condition, when one of the adult females bent right down to a drinking posture and when she stood up she had the skeletal remains of an animal about the size of a springbok dangling from her mouth. She had one of the hind legs in her mouth and was chewing away contentedly. The skeleton still had bits of hide and sinew holding it together so it swung around quite a bit which attracted the rest of the herd and within minutes there was a scuffle on the go to wrestle the prize away from the finder. If I didn't have this on film I wouldn't be writing about it as it's certainly not an everyday sighting. The tussle went on till the skeleton broke up and they all had their own pieces of bone to chew on. The giraffe's daily calcium and phosphorous requirements to sustain growth are large, and it is known that they get their calcium from what they browse on but the source of phosphorus is unknown. So it could be that they are deriving mineral benefits from chewing bones and not just having fun.

CHAPTER 35

TRAGEDY

At every available opportunity the animals are monitored by a member of the wildlife department. A young giraffe had been killed by the pride females and they were spotted feeding alone on the carcass. They had obviously done the hard work and were enjoying the tender bits before the rest of the pride arrived to join in. The pride stayed feeding on the carcass for three days and when there was literally nothing left of it they moved off. As you can imagine after three days of feeding all of the pride were as fat as ticks and in excellent health.

The next day the two white males were seen striding down the runway which, as the crow flies, was about six kilometres away from the carcass, which was no distance at all in lion terms. The next morning the whole pride was seen just north of the dam wall and all still in good condition. For the rest of the week only the pride females were seen. This got some alarm bells ringing and Scott went out to get a positive sighting of the males. The distance from the dam wall to where he found a signal of one of the males was about a kilometre. Only one signal was being received and this in itself was unusual as these two were inseparable. Both of the males had collars for tracking purposes.

The signal was coming from a very dense acacia thicket and he needed to get a visual on the lion. At the best of times it's not a good idea to go into thick bush knowing that there is a lion in it, but even worse if it's an acacia thicket because the long white thorns puncture human flesh with ease. Scott was worried because the signal hadn't moved and he made the decision to venture in. I can only imagine his nerves when he started to get close to the signal. Approaching a lion in the open is one thing but in a thicket it stretches your nerves to breaking point. After what must have felt like an eternity, Scott saw a white shape in the thicket. It wasn't moving and he observed for a few minutes but nothing stirred. He then started making noises but there was still no movement; at this point he decided to advance and it was then that he came across the dead body of the one male white lion. I can only surmise what he felt like – this was an animal that he had been tracking virtually every day for over a year.

The report came through and the team jumped into action as it was imperative to account for the entire pride as soon as possible. Sadly, it didn't take long to find the other male. He was also dead, about a kilometre east of the first one. The same situation, he was situated in a very dense acacia thicket. There were no marks on either of the bodies to suggest fighting or a bullet wound and it appeared that they died from some form of poisoning. It was imperative to get toxicological samples away quickly and to get an autopsy done as soon as possible. The carcasses were already in an advanced stage of deterioration and what was a very worrying fact was that the bodies had been scavenged on by other predators. The rest of the pride was found and they were all in the best of health which was a great relief in a very bad moment for the reserve. The autopsies revealed nothing conclusive, so the team had to wait for the toxicological reports.

To say that the reserve was turned upside down would be an understatement and I can only imagine how Paul felt, having to report this incident to the new bosses. Here was the hope of the white lion introduction programme's main gene pool gone! And not just the one male, but both simultaneously under the strangest of circumstances. Speculation was rife and has actually never ended because when the toxicological reports were received there was nothing in them that could tell us anything new at all. No toxins were found and I surmise that they had already degenerated to a point of non-recognition by the time they were harvested. To this day anything that anybody can say is purely speculation, and we won't ever know exactly what happened to these two incredible white lions who were truly the first to be free-roaming in the true sense of the word.

Once again decisions had to be made about the direction in which the programme was to go forward. The reserve was now getting busier and busier with guest numbers increasing all the time. The pride then consisted of two adult tawny lionesses and two sub-adult white females. The fact that quite a lot of mating had been going on between the white males and the tawny females gave the team some hope that there was a possibility of some progeny on the way. The loss of the two males was a terrible blow to all of the staff on the reserve; the sight of these two striding out across the veld had been an absolute joy. What was really heart-rending was that we weren't going to see that sight again. The fact that they were truly a free-roaming and surviving coalition white lion brotherhood made the loss that much greater.

That the programme had got to where it was is a reflection of the dedication of the team on the reserve and the foresight of Gaston Savoi in being able to visualise that it could happen. It was going to be a very tough road ahead to get the programme back to where it was before the loss of the two males. An aspect of the loss that was very difficult to measure was the guest experience. Male lions are a regal presence and tourists to our country have expectations of seeing a big male lion. For those who'd had the privilege of seeing the two adult male white lions free out in the wild, they were truly blessed. I've thought long and hard about what could have happened and, of course, what I now write is pure conjecture.

But knowing that most predators are not averse to scavenging, I think these two, who like most male lions relied on the females to supply their food requirements, found a carcass in the area near the dam that had built up some kind of toxin in it. Before the rest of the pride could consume whatever it was, the two males ate it just like they would normally do. I suspect it was a small animal or else the others would have had a chance of eating a portion of it, and this is why the rest of the pride didn't suffer the same fate. Knowing the area around the dam quite well it could have been anything from a hyrax to a terrapin, and then there is also the possibility that they were killed by a snake; however, in my opinion this is highly unlikely. As difficult as it is one must sometimes just accept that nobody will ever really know the answer.

Drought is a harsh reminder that nature is still in charge. After six years of above average rain the district entered into a dry cycle. Drought is a reality and the wildlife manager who does not manage his reserve for dry times is in for a very big wake-up call. Water is normally still available to the game from springs and man-made water sources, but in a drought most animals don't suffer from thirst but from hunger. Water holes have a distinct effect on the vegetation surrounding it. If you have a spring, the grazing and browsing effect is noticeable for up to five kilometres away. So it is very important to have water for the animals either naturally in the form of springs and rivers or in man-made water holes, but the vegetation must be considered before they are created. In the case of the Klein Karoo, the vegetation is incredibly hardy and resilient to drought but its makeup is such that it's highly vulnerable to overgrazing and trampling. For example, there are two spoor tracks that were created over 200 years ago by wagons, that are still visible to this day. Once the vegetation has been mechanically disturbed it is a permanent scar. The effects of goat and sheep farming will be visible for centuries to come.

CHAPTER 36

The Aftermath

After the death of the two white males, things went into limbo for a short while. There was talk of bringing in an habituated male but fortunately that decision was put aside. The team noticed that one of the white females was showing enlarged nipples and then she disappeared, which is a sure sign of an imminent birth. As mentioned before, lions literally only show their pregnancy in the last ten days of gestation.

After she had been away from the pride for about two weeks the team started to search for her with the telemetry. She was high up in a protected canyon when the signal was found and this was good news as her cubs would be safe in that environment. After three weeks or so she was seen with the pride – she didn't look good at all and was decidedly thin. There were no cubs with her as they would have been too young at that stage. The next question of course was: had she had any cubs and if so, where were they? For three days she remained with the pride and didn't return to where the team suspected she had given birth.

A normal behaviour pattern would have been to go back to the den to feed the cubs but because she was in such poor condition and she hadn't left the pride in three days, the decision was made to go and have a look. Making sure that she was still with the pride, a substantial distance from the kloof, the team started the search. Sadly, she had abandoned her two cubs. This is not an unusual occurrence for a first-time mother. However, these two little ones were still alive. Having the loss of the males weighing heavily on everybody's mind, it probably influenced the team's judgement and a decision was made to hand-rear the cubs. Fortunately, Shamwari has not only the facilities for hand rearing, but also staff who are experienced in hand-rearing orphaned animals, including lions. The hand-rearing was successful and the completely starved cubs not only survived but flourished. They had been fed by a dedicated team who wore white clothes and did everything possible not to imprint them.

While this was going on, the other white female left the pride and disappeared into the mountains. The plan for the orphan cubs was to reintroduce them to their mother in a controlled boma environment who would in turn introduce them to the pride on release, with the cubs still being young enough to not be a threat to any of the pride members.

Another decision that was to have a great impact was made by the powers that be. That same white female whose cubs were being hand-reared, was darted and put into an introduction boma with a young adult tawny male. As soon as they had settled down and were comfortable with each other through the fence, they were introduced. They settled down well and were comfortable with each other and over a few weeks there was no stress whatsoever between the two.

As soon as the hand-reared cubs were weaned, the idea was to put the two cubs into the boma next door and monitor what happened. After they had gained weight and were on solids they were brought back to Sanbona and put into the next door boma. At first there was a bit of tension which is understandable, but within days the female was spending all her time lying next to the cubs. I'm not sure if she had any idea that they were hers, but they certainly got on very well and played through the fence. After a while the male joined in and it started to look as if there was a chance of a successful bonding. The decision was made to do the introduction and long-term sedatives were given to the adults. For the first two weeks all went exceedingly well; there was a feeling of expectation that the lions would bond and we could get them back into their natural environment. Sadly, this was not to be and after a month together the worst case scenario happened. The male was taken back to where it came from and the female was released back into the reserve where she very quickly joined the pride.

It was an up and down time. It seemed that at every turn the project was doomed to failure and it was only when the second female came out of the mountains, followed by two white cubs that there was a glimmer of hope. She introduced them to the pride and they were accepted by all. So the reserve now had the two tawny females and the white female that had abandoned her first litter and the second white female with two white cubs. Luckily things settled down and there was no aggression within the pride at all.

However, a male lion was missing from the mix which was not a good situation to be in. Once the sex of the cubs was ascertained, some long-term planning would have to begin.

CHAPTER 37

More Threats Looming

Lions are facing a relatively new threat, as if they didn't have enough problems already! The Chinese are now making lion bone wine. There are fewer than 4,000 wild tigers left and only recently has the commercial trade in tiger parts been prohibited under international law. This prohibition has not stopped the slaughter and unbelievable cruelty meted out to some captive tigers, but has raised the price of the body parts of tigers on the black market. Because the Asians and traditional oriental medicine have denuded the tiger population, they are now turning to lion bone wine as a legal substitute for tiger bone wine. Asian consumers are being hoodwinked and may not even know this because the lion bone wine is frequently sold in tiger-shaped and labelled bottles.

Conservationists have been warning officials for years about the increased trade in lion parts; it has been going on for a long time and is increasing all the time. However, it has only recently enjoyed any publicity because certain criminal elements in the East and in South Africa have been caught smuggling lion bones. Eastern wines, made from tiger bones, have been around for centuries and this initially led to the elimination of most wild tiger populations.

Tiger bone wine is made by steeping tiger bones in huge vats of potent 38 per cent-proof rice wine. It is one of the most expensive and sought-after Chinese traditional medicines, believed to bestow the tiger's power and strength upon the person who drinks it. It's very popular among wealthy, middle-aged Asian men including, reportedly, some of the Communist Party's senior officials and is said to have been used by modern China's founder, Chairman Mao Tse-Tung himself, in the absolutely stupid belief that it counters the effects of ageing and boosts a flagging sex drive. Tigers are now very scarce because of the all-out war that these senseless idiots have declared on these magnificent creatures. The scarcity of tigers has driven the price of a single bottle of tiger bone wine to over $5,000. As well as being a supposed medicinal remedy, it is also a prestigious drink, shared between men at high-level political or business meetings, or drunk at lavish parties. I can only say that rhino horn, tiger and lion bones are not medicine in any form whatsoever and any person believing the contrary, is seriously unintelligent. I find it very difficult to believe that these so-called educated people are so incredibly stupid!

History has been exceedingly cruel to the tiger in China. Aside from the threat of hunting and poaching, in 1959 Chairman Mao decided to declare the tiger 'an enemy of the people.' Can you even begin to understand just what a self-serving imbecile this man was? It's really difficult to believe how incredibly dense and irrational some of our world leaders are. Sadly, to this day, politicians around the world are still being voted into power to destroy what is left of our planet. Half a century later, as the trade in tiger bones puts a huge price on the head of the remaining few wild tigers, the fate of the tiger in China is sealed. By the next Year of the Tiger (2022) China will not have any wild tigers left and I wonder how proud the people involved in this trade will be when they have to explain to their children what happened to the wild tigers. If you are repulsed by the canned lion hunting industry, then you are going to be doubly appalled at the tiger breeding facilities in China. They make the canned lion industry look like a Bambi outfit.

I had no idea how cruel the Chinese people were until I started researching the tiger bone wine industry. Go and look up the Xiongsen Tiger and Bear 'Park' in Guilan, and it will give you an indication of the abhorrent conditions that the Chinese allow their tigers to live in. According to researchers, most of the tigers are fed just enough for them not to die of starvation. And as if that's not enough, they have a family special at the park where Chinese families are invited to watch the tigers catch domestic animals on a concrete floor. In this modern world that we live in, it's now time for countries and individuals to take responsibility for their actions. The abuse of animals must be stopped in its tracks and the use of animal products in medicine must be outlawed and policed till it stops. I really find it incomprehensible that so-called intelligent people in this day and age still believe that tiger meat and tiger bone wine will slow the ageing process and enhance their sex lives!

Sadly, it isn't just the tigers that are in big trouble. These simple dimwits are also extracting bile from live bears to make throat lozenges, tea, wine and shampoo. They even use squirrel and bat faeces in their so-called medicines. Now we have these imbeciles taking lion bones to supplement the dwindling stocks of tiger bones. I feel strongly that before one has the right to criticise, which everybody should do, an alternative or possible solution to what you're criticising should be available. Many people ask me about the future of our wild areas and our nature reserves and my answer is very simple. Until we stop the rampant human expansion as soon as possible, we are going to lose our wildlife. The population growth must stop and, in fact, be reversed to the extent that we halve the existing population of the world over the next 100 years or so.

Our planet's resources are at breaking point; water is becoming a scarce commodity, and we are clogging up the air with the most terrible poisons. All of you reading this know exactly what the scenario is and we need more voices out there to stop this downhill slide into total annihilation. Now, just to rub it in our faces, our corruption-riddled ruling party has decided to give the go ahead for fracking exploration in the Karoo! If anybody can show me any benefit that fracking has brought about to mankind or the environment, please show me. The only people who benefit are the corrupt politicians and gas companies. Too many people and not enough resources; reverse the population growth and we shall have enough resources. That's my rant over for now in this book.

A new lodge named Dwyka has been built in the northern sector of Sanbona. It is a tented camp built in a place that we called Bobbejaan Krans, which means Baboon Cliff. The name Dwyka comes from the geological formation that was laid down over 300 million years ago. It was at a time when there was an ice age in the Karoo. A massive glacier covered the Karoo and the tillite soils are derived from the Dwyka sediments. These soils are hardy, stony clay-type soils which are acidic and drain poorly. It is a truly beautiful part of the reserve and I have had the privilege of sleeping out in the open under the stars there before there was any thought of it becoming a camp. There is a black eagle nest there that I've monitored over the years, and from which I've observed these raptors. The camp has been built in the most 'green' method possible, from the location of each of the tented bedrooms to the waste water. Every detail has been cleverly planned and implemented. Solar water heaters are used throughout the camp and this hot water is even used for the underfloor heating in winter. As hot as Sanbona becomes in summer, so bitterly cold does it get in winter, especially if there is snow on the mountains nearby and a breeze blows from that direction. Minus temperatures in the south are not uncommon but fortunately, by 10 am it's normally warm and clear. I was asked by Paul, who is now in charge of wildlife on the reserve, to assist the team in confirming all of the rock art and historical sites on the reserve. Liesl, Keir and I spent three days clambering around the gorges and kloofs confirming known places and found quite a few that have not been seen for centuries. The reserve is special in so many ways, with some of the places so remote that it is doubtful whether they will ever be explored.

CHAPTER 38

RETIREMENT

The reserve is now owned by Dubai World. Adrian Gardiner sold it to them and stayed on for a period, but is now no longer involved. Some of the senior staff are still in place and have been absorbed into the new company. In 2011, I filmed a documentary on the rock art of southern Africa and had permission from the new directors to film the rock art on Sanbona. It was wonderful to see and film the unique paintings that were left by the San/Bushmen. Of course, it was also good to see old friends and some of the animals that had been part of my life for so many years.

I did not set out to write a book. I started writing an article for a magazine, and it was only after I had reached 30 pages without covering half the story that I realised I was on my way to writing a book. Simmy has been encouraging me to do this for years, so it is actually her idea. The time that I spent on the reserve will be with me forever – the people, the animals, the whole environment and, of course, the lions. To this day I'm still reluctant to say which is my favourite animal species, but the lions have got to be up there in the top ten. Speak to any lion biologist about their chosen field of study and I can assure you they'll tell you that lions are possibly the most complicated, unpredictable, bloodthirsty, tender and wonderful (all at the same time) animals on this planet. They commit murder among their own one day and the next day are the most gentle and sensitive creatures in the world. I have witnessed a male kill its own cub. I suspect it was thinking that maybe the cub would be a challenge to him at a later stage, or possibly he didn't realise that it was his own. The term 'predictably unpredictable' is about the best way of summing up lion behaviour.

The reserve has now been up and running for 14 years and is in very capable hands. This, of course, is not the end of the story – it's just the end of the book. However, I have decided to follow the story from my own perspective and will do a follow-up on a yearly basis to relate what is happening, if the new owners allow it. There have been certain aspects that I have intentionally left out, not as a form of censorship, but as a way of protecting the team and the animals from the disgusting behaviour of the Asian medicine healers who, by their incredibly stupid belief that body parts of certain wild animals will cure various ailments, have put human and animal life in jeopardy.

As for the lions, well ... I have just visited the reserve and had the chance to catch up with Paul and Karlien and the team. The programme has had so many ups and downs and massive setbacks, but it is still on the go and I am convinced that eventually what we started all those years ago, will see the light of day.

The situation with the lions as of October 2012 is that there is a pride of five lions made up of two female tawny lions and one adult white female with her two white cubs, one male and one female. All three of the white lions were born and raised in the wild by wild lions with no interference from man whatsoever. This is incredible if you look back and remember all of the setbacks over the past ten years. The future is always unsure but what is certain is that the young white male cub must eventually be removed from the pride, if possible, to be introduced to a pride of females as he has never had any human interference. Then, if possible, a tawny lion should be introduced to the remaining females. This will give a new set of genes and the possibility of cubs being born, not white, but with the gene in them. When this occurs the project will have come of age and the progeny will have the opportunity of spreading those genes all over Africa.

Hindsight is an exact art. And yes, if I had the chance to do it all again, I would do some things differently, but not much – just some fine-tuning along the way. What made it all possible was Gaston Savoi's vision and a team that worked incredibly hard to try and bring it to reality. A truly great crew put their heads down and did the job with patience, diligence, dedication and professionalism, but mostly from hearts filled with love and passion. They did this, at times, under enormous pressure from many directions.

I have a strange habit in that whenever I travel, I always take one of my old pairs of shoes with me. And if the place that I travel to is special and I want to return to that place, I bury that pair of shoes there. Sanbona is so very special to me that there are two pairs hidden on the reserve and I am very happy to say that I've been able to visit both spots on a few occasions. Just to be in that part of the world and breathe in the beauty and splendour of the place does wonders for my being.

Having been there from the outset, I've been privileged to be part of a team that has taken the reserve from an overgrazed and incorrectly-utilised sheep and goat farming area with hundreds of tons of junk spread around, to a beautiful reserve with a conservation ethos that is admirable. My advice to the new owners is simple – look after the team that you have, treasure them because they love Sanbona and will do anything to see it thrive. Listen to them because they know what they're doing.

As for the founding pair, Jabulani and Queen, they have a beautiful double-boma retirement centre which gives them the opportunity of looking out over Sanbona and seeing the world go by. They have a delivery of fine-quality venison on a regular basis, compliments of the wildlife team. Queen has been sterilised to prevent any further breeding and is in beautiful condition. She is the founding mother and it is her progeny that have brought the first truly wild-born white lions on to the reserve. Jabulani is a massive male lion that has never had to fight for his pride or place, so he has a scar-free face which is highly unusual for his species. He is still my handsome boy, still amazingly inquisitive and has that cheeky streak in him that he had when he arrived all those years ago.

The two together are striking for many reasons, but then – I am possibly very slightly biased.

To anybody wanting to create a wilderness area or wildlife reserve in southern Africa, please feel free to contact me.

Note the pale lines under the eyes of both the white and tawny lions that reflect any available light into the eye, this is one of the factors that enhances a lions already incredibly good night vision.

Made in the USA
Monee, IL
04 November 2026

67386717R00154